Learning
Microeconometrics
with R

Chapman & Hall/CRC
The R Series

Series Editors
John M. Chambers, Department of Statistics, Stanford University, California, USA
Torsten Hothorn, Division of Biostatistics, University of Zurich, Switzerland
Duncan Temple Lang, Department of Statistics, University of California, Davis, USA
Hadley Wickham, RStudio, Boston, Massachusetts, USA

Recently Published Titles

Advanced R, Second Edition
Hadley Wickham

Dose Response Analysis Using R
Christian Ritz, Signe Marie Jensen, Daniel Gerhard, Jens Carl Streibig

Distributions for Modelling Location, Scale, and Shape
Using GAMLSS in R
Robert A. Rigby , Mikis D. Stasinopoulos, Gillian Z. Heller and Fernanda De Bastiani

Hands-On Machine Learning with R
Bradley Boehmke and Brandon Greenwell

Statistical Inference via Data Science
A ModernDive into R and the Tidyverse
Chester Ismay and Albert Y. Kim

Reproducible Research with R and RStudio, Third Edition
Christopher Gandrud

Interactive Web-Based Data Visualization with R, plotly, and shiny
Carson Sievert

Learn R: As a Language
Pedro J. Aphalo

Using R for Modelling and Quantitative Methods in Fisheries
Malcolm Haddon

R For Political Data Science: A Practical Guide
Francisco Urdinez and Andres Cruz
R Markdown Cookbook
Yihui Xie, Christophe Dervieux, and Emily Riederer

Learning Microeconometrics with R
Christopher P. Adams

R for Conservation and Development Projects
A Primer for Practitioners
Nathan Whitmore

For more information about this series, please visit: https://www.crcpress.com/
Chapman--HallCRC-The-R-Series/book-series/CRCTHERSER

Learning Microeconometrics with R

Christopher P. Adams

CRC Press
Taylor & Francis Group
Boca Raton London New York

CRC Press is an imprint of the
Taylor & Francis Group, an **informa** business

A CHAPMAN & HALL BOOK

First edition published 2021
by CRC Press
2 Park Square, Milton Park, Abingdon, Oxon, OX14 4RN

and by CRC Press
6000 Broken Sound Parkway NW, Suite 300, Boca Raton, FL 33487-2742

British Library Cataloguing-in-Publication Data
A catalogue record for this book is available from the British Library

Library of Congress Cataloging-in-Publication Data
[Insert LoC Data here when available]

ISBN: 9780367255381 (hbk)
ISBN: 9780429288333 (ebk)

Typeset in CMR10
by KnowledgeWorks Global Ltd.

To Deena and CJ.

Contents

Introduction

The Intern

You have been hired as a summer intern for a right-of-center think tank in Washington DC. It is going to be a great summer! You will play softball on the Mall. Go to Nats games. Hang out with friends interning on the Hill. And melt onto the sidewalk when you commute to work.

First Day

The think tank is arguing against a federal increase in the minimum wage. You have been asked to predict what will happen if the minimum wage increases from $7.25 to $15.00.

You have a vague memory of a discussion of the minimum wage in your Econ 101 class. To refresh your memory you google "minimum wage Khan Academy." You listen to Sol explain that the equilibrium wage is $6 per hour and workers work 22 million hours per month. Sol shows that a minimum wage of $7 leads to 2 million hours of unemployment and $1 million of output per month lost in the economy. This seems straightforward.

But what actually happens in the real world? Your supervisor suggests looking up minimum wages for each state and state level unemployment levels from the Bureau of Labor Statistics (`bls.gov`). She says that different states have changed their minimum wage over time and a number of states have minimum wages that are above $7.25, although none as high as $15.

You download the data on each state's current minimum wage and unemployment rate. You put everything in a spreadsheet. A fellow intern shows you how to save it as a "csv" file. He says this will allow importing the data into **R**, which is the statistical language of choice at your think tank.

You then download **R** and `RStudio` (the IDE you are told, whatever that is). Your colleague shows you how to get set up. He shows you how to open up `RStudio` and then create a new script file. You call the file "minwage.R" and save it to the minimum wage folder where you have the data set. He then tells you to go to "Session > Set Working Directory > To Source File Location." "Trust me. It makes coding easier," he says.

Now you are ready to write your first line of code.

```
> x <- read.csv("minimum wage.csv", as.is = TRUE)
```

Your colleague explains that `read.csv` will import the data set that you created. The data set is simply called **x**. He explains that you must use the **assign** `<-` arrow. You ask why. He shrugs, "that was what I was told when I started." Also, he says you should probably write `as.is = TRUE` because **R** has a habit of changing numbers to characters and other, even stranger, things.

You click **Run**. It worked! The letter x appears in the Global Environment. You click on it. A tab with your data appears.

You want to calculate the relationship between the minimum wage and unemployment. You want to run a **regression**.[1] You ask your cubicle neighbor how to do that. She tells you to write the following.

```
> lm1 <- lm(x$Unemployment.Rate ~ x$Minimum.Wage)
```

You ask about the whole thing with `<-`. Your neighbor says that you must do it that way but refuses to explain why.

You write out the code and hit **Run**. Nothing happens. Actually, `lm1` appears in the box in the upper right of the screen. Apparently it is a `List of 12`. You were hoping to see a table with regression results and t-statistics. But nothing. You ask for help from your neighbor. She rolls her eyes. "No. You just created an object called `lm1`. To look at it, use **summary**."

```
> summary(lm1)[4]
```

```
$coefficients
                 Estimate Std. Error    t value   Pr(>|t|)
(Intercept)    3.49960275 0.32494078 10.7699709 1.62017e-14
x$Minimum.Wage 0.01743224 0.03733321  0.4669365 6.42615e-01
```

Cool. You got what you were looking for. The minimum wage increases unemployment! It increases it by 0.01743. You wonder what that means. Another intern comes by, looks at what you did, and then types the following code on your computer. He leaves with a "you're welcome." Boy, is that guy annoying.

```
> a <- (15-7.25)*lm1$coefficients[2]
> a/mean(x$Unemployment.Rate)
```

```
x$Minimum.Wage
   0.03710335
```

Your neighbor explains it all. You want to know what happens to unemployment when the minimum wage increase from $7.25 to $15. The second coefficient states that amount. Then you can put it in percentage terms relative to the current unemployment rate.

---

[1]We use the term regression to refer to a function that summarizes the relationship in the data.

You go back to your supervisor. You say you found that a minimum wage increase to $15 would increase the unemployment rate by four percent. "The unemployment rate would go to 8%!" she exclaims. No, no, no. You clarify that it would increase by 4 percent not 4 percentage points. From say 4% to 4.15%. "Oh. Still that is a big increase." Then she says, "But are you sure? How can you tell what will happen in states that haven't changed their minimum wage?" You respond accurately. "I don't know." As you ponder this, you notice everyone is getting dressed for softball.

Second Day

On your way into the building you run into your supervisor. You explain how you were able to beat CBO's Dynamic Scorers by just one run. She congratulates you and says, "You should plot the relationship between the minimum wage and unemployment."

After some coffee and some googling you find code to do what your supervisor suggested.

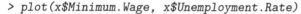

```
> plot(x$Minimum.Wage, x$Unemployment.Rate)
```

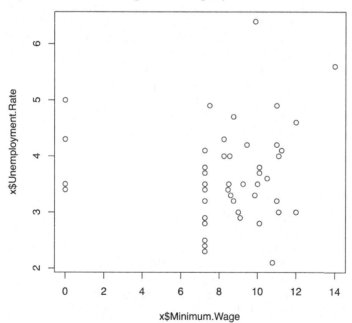

The annoying guy from yesterday breezes by and says. "Oh no. Looks like it is censored. You will need to use a **Tobit**." At this, someone a few cubicles away pops up like a meerkat. He says, "No. Don't use a Tobit, use a **Heckit**. The data is probably selected." Then he is gone.

What the heck is a Heckit? What the heck is a Tobit? What the heck is a meerkat?

The Book

The book is designed to help our fictitious intern hero survive the summer in DC.

What Does it Cover?

The book is based on a course I have taught at Johns Hopkins University as part of their Masters of Applied Economics. The book and the course aim to provide an introduction to applied microeconometrics. The goal is for the reader to have competence in using standard tools of microeconometrics including **ordinary least squares** (OLS), **instrumental variables** (IV), **probits** and **logits**. Not only should you be able to understand how these models work, but more importantly, you should be able to understand when they don't.

In addition to these standard models, the book and the course introduce important models commonly used in microeconometrics. These include the Tobit and the Heckman selection model (Heckit). The book introduces approaches that have become common in my sub field of empirical industrial organization. These approaches center around the analysis of **games**. That is, situations where a small number of individuals or firms interact strategically.

Lastly, the book introduces some new techniques that have been developed to analyze **panel data** models and other situations where an object of interest is measured repeatedly. It discusses **difference in difference** as well as **stochastic controls**. It touches on how machine learning techniques can be applied in microeconometrics. The book also introduces to a broader economic audience ideas of Harvard statistician, Herbert Robbins, called **empirical Bayesian** analysis.

What is the Approach?

The book teaches microeconometrics through **R**. It is not primarily aimed at teaching **R**. Rather, it is primarily aimed at teaching microeconometrics. This idea of using computer programming as a tool of instruction goes back to at least Seymour Papert and MIT's AI lab in the 1970s.[2] South African-born Papert helped develop a programming language called Logo. The goal of Logo was to teach mathematics by programming how a **turtle** moves around the screen. You may have used one of the offspring of Logo, such as Scratch or Lego Mindstorms.

I learned math through Logo. When I was a pre-teen, Logo became available on the personal computer, the Apple II. My parents taught me to program in Logo and I learned a number of geometric concepts such as Euclidean distance

---

[2]https://el.media.mit.edu/logo-foundation/

and the Pythagorean theorem by programming up a model of how a moth circled a light source.

The book uses Papert's ideas to teach microeconometrics. You will learn the math of the **estimator** and then how to program up that estimator.[3] The book makes particular use of the computer's ability to simulate data. This allows us to compare our estimates to what we know to be the true values. In some cases these simulations illustrate that our estimator is correct; in others, the simulations help us to understand why our estimator is incorrect. Testing models on simulated data has the additional benefit of allowing you to check your programming.

The book is written in `RStudio` using `Sweave`. `Sweave` allows LATEX to be integrated into **R**. LATEX is a free type-setting language that is designed for writing math. Almost all the code that is used in the book is actually presented in the book. On occasion it is more practical to create a data set outside the book. In those cases, the data and the code that created the data are available here `https://sites.google.com/view/microeconometricswithr/table-of-contents`. In a couple of other cases, the preferred code does not produce nice output for the book. I have highlighted those cases in the text. I also hide repetitive code. For the most part, the coding in the book is in base **R**. The book makes little use of packages. This shows you the underlying code and illustrates the econometrics. That said, there are a few packages that I really like, including `stargazer` and `xtable` which both make nice tables in LATEX.

What are POs, DAGs, and Do Operators?

POs, DAGs and Do Operators sound like school-yard put-downs, but they form the core of the book's approach to econometrics. This approach is heavily influenced by Northwestern econometrician, Charles Manski. I was lucky enough to have Chuck teach me econometrics in my first year of graduate school. It was a mind-altering experience. I had taken a number of econometrics classes as an undergraduate. I thought much of it was bunk. Manski said much of what you learned in standard econometrics classes was bunk. Manski gave meat to the bones of my queasiness with the econometrics I had been taught.

The book focuses on the question of **identification**. Does the algorithm estimate the parameter we want to know? Other books spend an inordinate amount of time on the accuracy of the parameter estimate or the best procedure for calculating the estimate. This book steps back and ask whether the procedure works at all. Can the data even answer the question? This seems to be fundamental to econometrics, yet is given short shrift in many presentations.

The book focuses on identifying the **causal effect**. What happens to the outcome of interest when the policy changes? What happens if college becomes

---

[3]An estimator is the mathematical method by which the relationship in the data is determined.

free? What happens if prices are increased? What happens if the federal minimum wage is increased?

To answer these causal questions, the book uses **directed acyclic graphs** (DAG) and **do operators**. The book, particularly the early chapters, relies on ideas of Israeli and UCLA computer scientist, Judea Pearl. DAGs help us to understand whether the parameter of interest can be identified from the data we have available. These diagrams can be very useful models of the **data generating process**.[4] Hopefully, it will be clear that DAGs are models, and as such, they highlight some important issues while suppressing others.

Pearl's do operator helps illustrate the old statistical chestnut, "correlation is not causality." Observing the unemployment rate for two different minimum wage laws in two different states is quite different from changing the minimum wage law for one state. In the first instance we observe a statistical quantity, the unemployment rate conditional on the minimum wage law. In the second case we are making a prediction, what will happen to the unemployment rate if the law is changed?

In some cases it is useful to illustrate these issues using the **potential outcome** (PO) model of former Harvard statistician, Donald Rubin. This model highlights the fundamental identification problem of statistics. We can never observe the difference in state unemployment rates for two different minimum wage laws. Sure, we can observe unemployment rates for two different states with different minimum wage laws. We can even observe the difference in unemployment rates for the same state before and after a change in the minimum wage law. However, we cannot observe the unemployment rate for the same state at the same time with two different minimum wage laws.

In addition, the PO model illustrates that causal effect is not single valued. A policy that encourages more people to attend college may allow many people to earn higher incomes, but it may not help all people. It is even possible that some people are made worse off by the policy. There is a distribution of causal effects.

What About the Real World?

The course I have taught at Hopkins is for a Masters of *Applied* Economics. I take the *applied* part of this seriously. The course and this book aim to show how to do microeconometrics. I have spent my career using data to answer actual policy questions. Did a realtor group's policies lead to higher prices for housing transactions?[5] Did Google's changes to the search results page harm competitors or help consumers?

The book presents interesting and important questions. One of the most

---

[4]Data generating process refers to where the data we use comes from. For example, in drug development, data on the effects of the drug come from double blind randomly assigned trials with a placebo control. This process has large implications for what information can be inferred from the data.

[5]http://www.opn.ca6.uscourts.gov/opinions.pdf/11a0084p-06.pdf

important is measuring "returns to schooling." What is the causal effect on income of having one more year of school? It is easy to see that people with college degrees earn more than those with high school diplomas. It is much harder to determine if a policy that encourages someone to finish college actually leads that person to earn more money. I throw lots of data, economic theory and statistical techniques at this question. Hopefully, by the end you will see how analysis of survey data with **ordinary least squares** (OLS), **instrumental variables** (IV), **Heckman selection** and **generalized method of moments** models helps us answer this question. You will also see how **mixture models** can be used to analyze comparisons of twins.

The book discusses important questions beyond returns to schooling. It discusses racism in mortgage lending. It discusses gender bias in labor market earnings. It discusses increasing the federal minimum wage. It discusses the effect of guns on crime. It even discusses punting on fourth down. I hope the book points you to new questions and new data to answer existing questions.

The book does not recommend policies. The government economist and founding Director of the Congressional Budget Office, Alice Rivlin, argued that it is extremely important to provide policy makers with objective analysis. In a memo to staff she said the following.[6]

> We are not to be advocates. As private citizens, we are entitled to our own views on the issues of the day, but as members of CBO, we are not to make recommendations, or characterize, even by implication, particular policy questions as good or bad, wise or unwise.

Economists in government, the private sector and the academy work on important policy questions. I believe that economists are most effective when they do not advocate for policy positions, but present objective analysis of the economics and the data. I hope that this book presents objective analysis of interesting policy questions and you have no idea whether I think particular policy positions are good or bad, wise or unwise.[7]

The Outline

The book's twelve chapters are broken into three parts based on the main approach to **identification**. The first part presents methods that rely on the existence of an experiment. This part includes chapters covering ordinary least squares, instrumental variables (IV), randomized controlled trials (RCTs) and Manski bounds. The second part presents methods that rely on economic theory to identify parameters of interest. This is often referred to as a **structural**

---

[6]https://www.cbo.gov/sites/default/files/Public_Policy_Issues_Memo_Rivlin_1976.pdf

[7]In almost all cases, you should punt on fourth down.

approach. These chapters discuss demand models and discrete estimators such as logits and probits, censored and selection models, non-parametric auction models and generalized method of moments (GMM). The third part presents methods that rely on the existence of repeated measurement in the data. These methods include difference in difference, fixed effects, synthetic controls and factor models.

Experiments

The first four chapters rely on experiments, broadly construed, to identify the causal effects of policies.

Chapter 1 introduces the work-horse algorithm of economics, ordinary least squares (OLS). This model is simple and quick to estimate and often produces reasonable results. The chapter illustrates how the model is able to disentangle the effects on the outcome of interest. OLS relies on strong assumptions. In particular, the model assumes that the policy variable of interest affects the outcome **independently** of any unobserved term.

Chapter 2 considers how additional observed characteristics improve our estimates. It shows when adding more **control variables** improves the estimation and when it produces garbage. The chapter discusses the problem of **multicollinearity**. It discusses an alternative to the standard approach based on the work of Judea Pearl. The chapter replicates the OLS model used in Card (1995) to estimate returns to schooling. The chapter uses a DAG and Pearl's approach to help determine whether there exists evidence of systematic racism in mortgage lending.

Chapter 3 introduces the **instrumental variables** model. This model allows the **independence assumption** to be weakened. The model allows the policy variable to be affected by unobserved characteristics that also determine the outcome. The chapter presents IV estimates of returns to schooling by replicating Card (1995). DAGs are used to illustrate and test the assumptions. The local average treatment effect (LATE) is proposed as an estimator when the researcher is unwilling to assume the treatment effects each person identically.

Chapter 4 considers formal experiments. The ideal randomized controlled trial allows the researcher to estimate the average effect of the policy variable. It also allows the researcher to bound the distribution of effects using **Kolmogorov bounds**. The method is used to bound the effect of commitment savings devices on increasing savings. The chapter presents Manski's **natural bounds** and discusses inference when the data does not come from ideal randomized controlled trials. It considers the problem of estimating the causal effect of guns on crime using variations in state gun laws.

Structural Estimation

The first four chapters consider questions and issues relevant to economics, but describe standard estimation methods. Chapters 5 to 9 use economic theory directly in the estimation methods.

Chapter 5 introduces **revealed preference**. The chapter shows how this idea is used to infer unobserved characteristics of individual economic actors. Berkeley econometrician, Dan McFadden, pioneered the idea of using economic theory in his analysis of how people would use the new (at the time) Bay Area Rapid Transit (BART) system. This chapter introduces standard tools of demand analysis including the logit and probit models. It takes these tools to the question of whether smaller US cities should invest in urban rail infrastructure.

Chapter 6 also uses revealed preference, but this time to analyze labor markets. Chicago's Jim Heckman shared the Nobel prize with Dan McFadden for their work on revealed preference. In McFadden's model you, the econometrician, do not observe the outcome from any choice, just the choice that was made. In Heckman's model you observe the outcome from the choice that was made, but not the outcome from the alternative. The chapter describes the related concepts of **censoring** and **selection**, as well as their model counterparts the Tobit and Heckit. The chapter uses these tools to analyze gender differences in wages and returns to schooling.

Chapter 7 returns to the question of estimating demand. This time it allows the price to be determined as the outcome of market interactions by a small number of firms. This chapter considers the modern approach to demand analysis developed by Yale economist, Steve Berry. This approach combines **game theory** with IV estimation. The estimator is used to determine the value of Apple Cinnamon Cheerios.

Chapter 8 uses **game theory** and the concept of a **mixed strategy Nash equilibrium** to reanalyze the work of Berkeley macroeconomist, David Romer. Romer used data on decision making in American football to argue that American football coaches are not rational. In particular, coaches may choose to punt too often on fourth down. Reanalysis with game theory finds the choice to punt to be generally in line with the predictions of economic theory. The chapter introduces the **generalized method of moments** (GMM) estimator developed by the University of Chicago's Nobel laureate Lars Peter Hansen.

Chapter 9 considers the application of game theory to auction models. The book considers the GPV model of first price (sealed bid) auctions and the Athey-Haile model of second price (English) auctions. GPV refers to the paper by Emmanuel Guerre, Isabelle Perrigne and Quang Vuong, *Optimal Nonparametric Estimation of First-Price Auctions* published in 2000. The paper promoted the idea that auctions, and structural models more generally, can be estimated in two steps. In the first step, standard statistical methods are used to estimate statistical parameters of the auction. In the second step, economic theory is used to back out the underlying policy parameters. For second-price auctions, the chapter presents the **order statistic** approach of Athey and Haile (2002). Stanford's Susan Athey and Yale's Phil Haile provide a method for analyzing auctions when only some of the information is available. In particular, they assume that the econometrician only knows the price and the number of bidders. These methods are used to analyze timber auctions and determine whether the US Forestry Service had legitimate concerns about collusion in the 1970s logging industry.

Repeated Measurement

Chapters 10 to 12 consider data with repeated measurement. Repeated measurement has two advantages. First, it allows the same individual to be observed facing two different policies. This suggests that we can measure the effect of the policy as the difference in observed outcomes. Second, repeated measurement allows the econometrician to infer unobserved differences between individuals. We can measure the value of a policy that affects different individuals differently.

Chapter 10 considers panel data models. Over the last 25 years, the difference in difference estimator has become one of the most used techniques in microeconometrics. The chapter covers difference in difference and the standard fixed effects model. These methods are used to analyze the impact of increasing the minimum wage. The chapter replicates David Card and Alan Krueger's famous work on the impact of increasing the minimum wage in New Jersey on restaurant employment. The chapter also measures the impact of the federal increase in the minimum wage that occurred in the late 2000s. The chapter follows the work of Princeton labor economist Janet Currie and uses fixed effects and panel data from the National Longitudinal Survey of Youth 1997 (Currie and Fallick, 1996).

Chapter 11 considers a more modern approach to panel data analysis. Instead of assuming that time has the same effect on everyone, the chapter considers various methods for creating **synthetic controls**. It introduces the approach of Abadie et al. (2010) as well as alternative approaches based on regression regularization and convex factor models. It discusses the benefits and costs of these approaches and compares them using NLSY97 to measure the impact of the federal increase in the minimum wage in the late 2000s.

Chapter 12 introduces mixture models. These models are used throughout microeconometrics, but they are particularly popular as a way to solve measurement error issues. The chapter explains how these models work. It shows that they can be identified when the econometrician observes at least two signals of the underlying data process of interest. The idea is illustrated estimating returns to schooling for twins. This is based on the work of Princeton labor economists Orley Ashenfelter, Alan Krueger and Cecilia Rouse. The chapter returns to the question of the effect of New Jersey's minimum wage increase on restaurant employment. The mixture model is used to suggest that the minimum wage increase *reduced* employment for small restaurants, consistent with economic theory.

Technical Appendices

The book has two technical appendices designed to help the reader to go into more depth on some issues that are not the focus of the book.

Appendix A presents statistical issues, including assessing the value of estimators using measures of **bias**, **consistency** and accuracy. It presents a discussion of the two main approaches to finding estimators, the **classical** method and the **Bayesian** method. It discusses standard classical ideas based on the **Central Limit Theorem** and a more recent innovation known as **bootstrapping**. The Bayesian discussion includes both standard ideas and Herbert Robbins' empirical Bayesian approach. Like the rest of the book, this appendix shows how you can use these ideas but also gives the reader some insight on why you would want to use them. The appendix uses the various approaches to ask whether John Paciorek was better than Babe Ruth.

Appendix B provides more discussion of **R** and various programming techniques. The appendix discusses how **R** is optimized for analysis of vectors, and the implications for using loops and optimization. The chapter discusses various objects that are used in **R**, basic syntax and commands as well as basic programming ideas including `if ()` `else`, `for ()` and `while ()` loops. The appendix discusses how matrices are handled in **R**. It also provides a brief introduction to optimization in **R**.

Notation

As you have seen above, the book uses particular fonts and symbols for various important things. It uses the symbol **R** to refer to the scripting language. It uses `typewriter font` to represent code in **R**. Initial mentions of an important term are in **bold face font**.

In discussing the data analysis it uses X to refer to some observed characteristic. In general, it uses X to refer to the policy variable of interest, Y to refer to the outcome, and U to refer to the unobserved characteristic. When discussing actual data, it uses x_i to refer to the observed characteristic for some individual i. It uses $\mathbf{x}$ to denote a vector of the x_i's. For matrices it uses $\mathbf{X}$ for a matrix and $\mathbf{X}'$ for the matrix transpose. A row of that matrix is $\mathbf{X}_i$ or $\mathbf{X}_i'$ to highlight that it is a row vector. Lastly for parameters of interest it uses Greek letters. For example, β generally refers to a vector of parameters, although in some cases it is a single parameter of interest, while $\hat{\beta}$ refers to the estimate of the parameter. An individual parameter is b.

Hello R World

To use this book you need to download **R** and RStudio on your computer. Both are free.

Download R and RStudio

First, download the appropriate version of RStudio here: `https://www.rstudio.com/products/rstudio/download/#download`. Then you can download the appropriate version of **R** here: `https://cran.rstudio.com/`.

Once you have the two programs downloaded and installed, open up RStudio. To open up a *script* go to "File > New File > R Script." You should have four windows: a script window, a console window, a global environment window, and a window with help, plots and other things.

Using the Console

Go to the console window and click on the >. Then type `print("Hello R World")` and hit enter. Remember to use the quotes. In general, **R** functions have the same basic syntax, `functionname` with parentheses, and some input inside the parentheses. Inputs in quotes are treated as text while inputs without quotes are treated as variables.

```
> print("Hello R World")
```

```
[1] "Hello R World"
```

Try something a little more complicated.

```
> a <- "Chris"  # or write your own name
> print(paste("Welcome",a,"to R World",sep=" "))
```

```
[1] "Welcome Chris to R World"
```

Here we are creating a variable called **a**. To define this variable we use the <- symbol which means "assign." It is possible to use = but that is generally frowned upon. I really don't know why it is done this way. However, when writing this out it is important to include the appropriate spaces. It should be `a <- "Chris"` rather than `a<-"Chris"`. Not having the correct spacing can lead to errors in your code. Note that # is used in **R** to "comment out" lines in codes. **R** does not read the line following the hash.

In **R** we can place one function inside another function. The function `paste` is used to join text and variables together. The input `sep = " "` is used to place a space between the elements that are being joined together. When placing one function inside another make sure to keep track of all of the parentheses. A common error is to have more or fewer closing parentheses than opening parentheses.

A Basic Script

In the script window name your script. I usually name the file something obvious like `Chris.R`. You can use your own name unless it is also Chris.

```
> # Chris.R
```

Note that this line is commented out, so it is does nothing. To actually name your file you need to go to "File > Save As" and save it to a folder. When I work with data, I save the file to the same folder as the data. I then go to "Session > Set Working Directory > To Source File Location." This sets the working directory to be the same as your data. It means that you can read and write to the folder without complex path names.[8]

Now you have the script set up. You can write into it.

```
> # Chris.R
>
> # setwd("WRITE NAME OF DIRECTORY HERE")
> # an alternative way to set working directory.
>
> # Import data
> x <- read.csv("minimum wage.csv", as.is = TRUE)
> # the data can be imported from here:
> # https://sites.google.com/view/microeconometricswithr/
> # table-of-contents
>
> # Summarize the data
> summary(x)
```

```
     State           Minimum.Wage      Unemployment.Rate
 Length:51         Min.   : 0.000    Min.   :2.100
 Class :character  1st Qu.: 7.250    1st Qu.:3.100
 Mode  :character  Median : 8.500    Median :3.500
                   Mean   : 8.121    Mean   :3.641
                   3rd Qu.:10.100    3rd Qu.:4.100
                   Max.   :14.000    Max.   :6.400
```

```
> # Run OLS
> lm1 <- lm(Unemployment.Rate ~ Minimum.Wage, data = x)
> summary(lm1)[4]
```

```
$coefficients
                Estimate Std. Error   t value      Pr(>|t|)
(Intercept)   3.49960275 0.32494078 10.7699709 1.62017e-14
Minimum.Wage  0.01743224 0.03733321  0.4669365 6.42615e-01
```

---

[8]However, be careful to set the working directory as you may end up putting files in places you didn't mean!

```
> # the 4th element provides a nice table.
```

To run this script, you can go to the "Run > Run All". Alternatively you can move the cursor to each line and hit `ctrl enter`. The first line imports the data. You can find the data at the associated website for the book. The data has three variables: the State, their June 2019 minimum wage, and their June 2019 unemployment rate. To see these variables, you can run `summary(x)`.

To run a standard regression use the `lm()` function. I call the object `lm1`. On the left-hand side of the tilde ($\sim$) is the variable we are trying to explain, `Unemployment.Rate`, and on the right-hand side is the `Minimum.Wage` variable. If we use the option `data = x`, we can just use the variable names in the formula for the `lm()` function.

The object, `lm1`, contains all the information about the regression. You can run `summary(lm1)` to get a standard regression table.[9]

Discussion and Further Reading

The book is a practical guide to microeconometrics. It is not a replacement for a good textbook such as Cameron and Trivedi (2005) or classics like Goldberger (1991) or Greene (2000). The book uses **R** to teach microeconometrics. It is a complement to other books on teaching **R**, particularly in the context of econometrics, such as Kleiber and Zeileis (2008). For more discussion of DAGs and do operators, I highly recommend Judea Pearl's *Book of Why* (Pearl and Mackenzie, 2018). To learn about programming in **R**, I highly recommend Kabacoff (2011). The book is written by a statistician working in a computer science department. Paarsh and Golyaev (2016) is an excellent companion to anyone starting out doing computer intensive economic research.

---

[9]Unfortunately, there are some issues with characters, so instead I just present the table of regression results, which is the 4th element of the list created by `summary()`.

Part I

Experiments

1

Ordinary Least Squares

1.1 Introduction

Ordinary least squares (OLS) is the work-horse model of microeconometrics. It is quite simple to estimate. It is straightforward to understand. It presents reasonable results in a wide variety of circumstances.

The chapter uses OLS to disentangle the effect of the policy variable from other unobserved characteristics that may be affecting the outcome of interest. The chapter describes how the model works and presents the two standard algorithms, one based on matrix algebra and the other based on **least squares**. It uses OLS on actual data to estimate the value of an additional year of high school or college.

1.2 Estimating the Causal Effect

You have been asked to evaluate a policy which would make public colleges free in the state of Vermont. The Green Mountain state is considering offering free college education to residents of the state. Your task is to determine whether more Vermonters will attend college and whether those additional attendees will be better off. Your boss narrows things down to the following question. Does an additional year of schooling **cause** people to earn more money?

The section uses simulated data to illustrate how **averaging** is used to disentangle the effect of the policy variable from the effect of the unobserved characteristics.

1.2.1 Graphing the Causal Effect

Your problem can be represented by the **causal graph** in Figure 1.1. The graph shows an arrow from X to Y and an arrow from U to Y. In the graph, X represents the policy variable. This is the variable that we are interested in changing. In the example, it represents years of schooling. The outcome of interest is represented by Y. Here, the outcome of interest is income. The arrow from X to Y represents the **causal effect** of X on Y. This means that a change

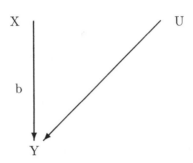

FIGURE 1.1
Causal graph.

in X will lead to a change in Y. The size of the causal effect is represented by b. The model states that if a policy encourages a person to gain an additional year of schooling (X), then the person's income (Y) increases by the amount b. Your task is to estimate b.

If the available data is represented by the causal graph in Figure 1.1, can it be used to estimate b? The graph shows there is an unobserved term U that is also affecting Y. U may represent unobserved characteristics of the individual or the place where they live. In the example, some people live in Burlington while others live in Northeast Kingdom. Can we disentangle the effect of X on Y and the effect of U on Y?

1.2.2 A Linear Causal Model

Consider a simple model of our problem. Individual i earns income y_i determined by their education level x_i and unobserved characteristics v_i. We assume that the relationship is a linear function.

$$y_i = a + bx_i + v_i \tag{1.1}$$

where a and b are the parameters that determine how much income individual i earns and how much of that is determined by their level of education.

Our goal is to estimate these parameters from the data we have.

1.2.3 Simulation of the Causal Effect

In the simulated data, we have a true linear relationship between x and y with an intercept of 2 and a slope of 3. The unobserved characteristic is distributed **standard normal** $(v_i \sim \mathcal{N}(0,1))$. This means that the unobserved characteristic is drawn from a normal distribution with mean 0 and standard deviation of 1. We want to estimate the value of b, which has a true value of 3.

```
> # Create a simulated data set
> set.seed(123456789)
> # use to get the exact same answer each time the code is run.
> # you need to set the seed each time you want to get the
> # same answer.
> N <- 100
> # Set N to 100, to represent the number of observations.
> a <- 2
> b <- 3 # model parameters of interest
> # Note the use of <- to mean "assign".
> x <- runif(N)
> # create a vector where the observed characteristic, x,
> # is drawn from a uniform distribution.
> u <- rnorm(N)
> # create a vector where the unobserved characteristic,
> # u is drawn from a standard normal distribution.
> y <- a + b*x + u  # create a vector y
> # * allows a single number to be multiplied through
> # the whole vector
> # + allows a single number to be added to the whole vector
> # or for two vectors of the same length to be added together.
```

A computer does not actually generate random numbers. It generates pseudo-random numbers. These numbers are derived from a distinct function. If you know the function and the current number, then you know exactly what the next number will be. The book takes advantage of this process by using the **R** function, `set.seed()` to generate exactly the same numbers every time.

Figure 1.2 presents the true relationship between X and Y as the solid line sloping up. The observed relationship is represented as the circles randomly spread out over the box. The plot suggests that we can take averages in order to determine the true relationship between X and Y.

1.2.4 Averaging to Estimate the Causal Effect

Figure 1.2 suggests that we can use averaging to disentangle the effect of X on Y from the effect of U on Y. Although the circles are spread out in a cloud, they follow a distinct upward slope. Moreover, the line that sits in the center of the cloud represents the true relationship. If we take values of X close to 0, then the average of Y will be near 2. If we look at values of X close to 1, the average of Y is near 5. That is, if we change the value of X by 1, the average value of Y increases by 3.

```
> mean(y[x > 0.95]) - mean(y[x < 0.05])
```

```
> plot(x, y)   # creates a simple plot
> abline(a = 2, b = 3)   # adds a linear function to the plot.
> # a - intercept, b - slope.
```

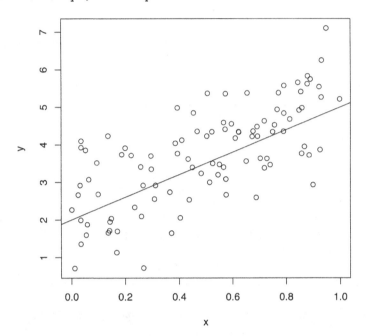

FIGURE 1.2
Plot of x and y with the true relationship represented by the line.

```
[1] 2.721387
```

```
> # mean takes an average
> # the logical expression inside the square brackets
> # creates an index for the elements of y where the logical
> # expression in x holds.
```

We can do this with the simulated data. In both cases, we are "close" but not actually equal to 0 or 1. The result is that we find a slope that is equal to 2.72, not 3.

We can disentangle the effect of X and U by averaging out the unobserved characteristic. By taking the difference in the average of Y calculated at two different values of X, we can determine how X affects the average value of Y. In essence, this is what OLS does.

1.2.5 Assumptions of the OLS Model

The major assumptions of the OLS model are that unobserved characteristics enter **independently** and **additively**. The first assumption states that conditional on observed characteristics (the X's), the unobserved characteristic (the U) has independent effects on the outcome of interest (Y). In Figure 1.1, this assumption is represented by the fact there is no arrow from U to X. In the econometrics literature, the assumption is given the awkward appellation, **unconfoundedness** (Imbens, 2010).

To understand the independence assumption, it is helpful to go back to the original problem. We are interested in determining the economic value of attending college. Our estimated model of the effect of schooling on income allows the unobserved characteristics to determine income. Importantly, the model does not allow unobserved characteristics to affect *both* schooling and income. The model does not allow students from wealthy families to be more likely to go to college and get a good job due to their family background.

The second assumption states that unobserved characteristics enter the model additively.[1] The implication is that the effect of the policy cannot vary with unobserved characteristics. Attending college increases everyone's income by the same amount. This assumption allows the effect of U and X to be disentangled using **averaging**. This assumption also allows the use of the quick and robust **least squares** algorithm.

The simulated data set satisfies these two assumptions. The unobserved characteristic, v, is drawn **independently** of x and it affects y **additively**.

1.3 Matrix Algebra of the OLS Model

The chapter presents two algorithms for solving the model and revealing the parameter estimates; the algebraic algorithm and the least squares algorithm.[2] This section uses matrix algebra to derive and program up the OLS estimator in **R**.[3]

---

[1] OLS can estimate models where the unobserved characteristic enters multiplicatively as this model is additive in logs.

[2] Under additional assumptions we can use other algorithms such as **maximum likelihood** or the **generalized method of moments**. See Chapter 5 and 8 respectively.

[3] For a matrix algebra refresher check out *Khan Academy* and Sol's excellent presentation of the basic concepts. Appendix B presents additional discussion of the use of matrices in **R**.

1.3.1   Standard Algebra of the OLS Model

Let's simplify the problem. Consider Equation (1.1) and let $a = 2$, so we only need to determine the parameter b.

$$b = \frac{y_i - 2 - v_i}{x_i} \tag{1.2}$$

The parameter of interest (b) is a function of both observed terms ($\{y_i, x_i\}$) and the unobserved term (v_i).

Equation (1.2) highlights two problems. The first problem is that the observed terms ($\{y_i, x_i\}$) are different for each person i, but Equation (1.2) states that b is exactly the same for each person. The second problem is that the unobserved term (v_i) is unobserved. We do not know what it is.

Luckily, we can "kill two birds with one stone." As Equation (1.2) must hold for each individual i in the data, we can determine b by averaging. Going back to the original equation (Equation (1.1)), we can take the average of both sides of the equation.

$$\frac{1}{N} \sum_{i=1}^{N} y_i = \frac{1}{N} \sum_{i=1}^{N} (2 + bx_i + v_i) \tag{1.3}$$

We use the notation $\sum_{i=1}^{N}$ to represent summing over N individuals in data. The Greek letter $\sum$ is capital sigma.[4]

Summing through on the right-hand side, we have the slightly simplified equation below.

$$\frac{1}{N} \sum_{i=1}^{N} y_i = 2 + b\frac{1}{N} \sum_{i=1}^{N} x_i + \frac{1}{N} \sum_{i=1}^{N} v_i$$

$$\tag{1.4}$$

or

$$\bar{y} = 2 + b\bar{x} + \bar{v}$$

where $\bar{y}$ denotes the sample average.

Dividing by $\bar{x}$ and rearranging to make the parameter of interest (b) as a function of the observed and unobserved terms:

$$b = \frac{\bar{y} - 2 - \bar{v}}{\bar{x}} \tag{1.5}$$

Unfortunately, we cannot actually solve this equation and determine b. The problem is that we still cannot observe the unobserved terms, the v_i's.

However, we can estimate b using Equation (1.5). The standard notation is $\hat{b}$.

$$\hat{b} = \frac{\bar{y} - 2}{\bar{x}} \tag{1.6}$$

---

[4]This is useful to remember if you are ever hungry in Athens because it is the first letter of the famous Greek sandwich, the souvlaki (Σουνλακι).

The estimate of b is a function of things we observe in the data. The good news is that we can calculate $\hat{b}$. The bad news is that we don't know if it is equal to b.

How close is our estimate to the true value of interest? How close is $\hat{b}$ to b? If we have a lot of data and it is reasonable to assume the mean of the unobserved terms is zero, then our estimate will be very close to the true value. If we assume that $\mathbb{E}(v_i) = 0$ then by the **Law of Large Numbers** if N is large, $\frac{1}{N}\sum_{i=1}^{N} v_i = \bar{v} = 0$ and our estimate is equal to the true value $(\hat{b} = b)$.[5]

1.3.2 Algebraic OLS Estimator in R

We can use the algebra presented above as pseudo-code for our first estimator in **R**.

```
> b_hat <- (mean(y) - 2)/mean(x)
> b_hat
```

```
[1] 3.459925
```

Why does this method not give a value closer to the true value? The method gives an estimate of 3.46, but the true value is 3. One reason may be that the sample size is not very large. You can test this by running the simulation and increasing N to 1,000 or 10,000. Another reason may be that we have simplified things to force the intercept to be 2. Any variation due to the unobserved term will be taken up by the slope parameter giving an inaccurate or **biased** estimate.[6]

1.3.3 Using Matrices

In general, of course, we do not know a and so we need to solve for both a and b. We use **matrix algebra** to solve this more complicated problem.

Equation (1.7) shows the equations representing the data. It represents a system of 100 linear equations.

$$
\begin{bmatrix} y_1 \\ y_2 \\ y_3 \\ y_4 \\ y_5 \\ \vdots \end{bmatrix} = \begin{bmatrix} 2 + 3x_1 + v_1 \\ 2 + 3x_2 + v_2 \\ 2 + 3x_3 + v_3 \\ 2 + 3x_4 + v_4 \\ 2 + 3x_5 + v_5 \\ \vdots \end{bmatrix} \qquad (1.7)
$$

---

[5]See discussion in Appendix A.

[6]An estimate is said to be biased if we take a large number of imaginary samples and the average imaginary estimate differs from the true value. This issue is discussed in detail in Appendix A.

Using matrix algebra, we can rewrite the system of equations.

$$
\begin{bmatrix} y_1 \\ y_2 \\ y_3 \\ y_4 \\ y_5 \\ \vdots \end{bmatrix} = \begin{bmatrix} 1 & x_1 \\ 1 & x_2 \\ 1 & x_3 \\ 1 & x_4 \\ 1 & x_5 \\ & \vdots \end{bmatrix} \begin{bmatrix} 2 \\ 3 \end{bmatrix} + \begin{bmatrix} v_1 \\ v_2 \\ v_3 \\ v_4 \\ v_5 \\ \vdots \end{bmatrix} \tag{1.8}
$$

Notice how matrix multiplication is done. Standard matrix multiplication follows the rule below.

$$
\begin{bmatrix} a & b \\ c & d \end{bmatrix} \begin{bmatrix} e & f \\ g & h \end{bmatrix} = \begin{bmatrix} ae + bg & af + bh \\ ce + dg & cf + dh \end{bmatrix} \tag{1.9}
$$

Check what happens if you rearrange the order of the matrices. Do you get the same answer?[7]

1.3.4   Multiplying Matrices in R

We now illustrate **matrix multiplication** with **R**.

```
> x1 = x[1:5] # only include elements 1 to 5.
> X1 = cbind(1,x1) # create a matrix with a columns of 1s
> # cbind means column bind -
> # it joins columns of the same length together.
> # It returns a matrix-like object.
> # Predict value of y using the model
> X1%*%c(2,3)

          [,1]
[1,] 4.079527
[2,] 4.018643
[3,] 3.961705
[4,] 4.156967
[5,] 4.764634

> # See how we can add and multiply vectors and numbers in R.
> # In R %*% represents standard matrix multiplication.
> # Note that R automatically assumes c(2,3) is a column vector
> # Compare to the true values
> y[1:5]

[1] 4.224902 4.219999 5.374130 3.378327 5.541021
```

In the simulated data we see the relationship between y and x. Why aren't the predicted values èqual to the true values?[8]

---

[7] No. In matrix algebra, order matters!

[8] They are not generally equal because we didn't include the unobserved term (v).

1.3.5 Matrix Estimator of OLS

It is a lot more compact to represent the system in Equation (1.7) with matrix notation.

$$\mathbf{y} = \mathbf{X}\beta + \upsilon \qquad (1.10)$$

where $\mathbf{y}$ is a 100×1 column vector of the outcome of interest y_i, $\mathbf{X}$ is a 100×2 rectangular matrix of the observed explanatory variables $\{1, x_i\}$, β is a 2×1 column vector of the model parameters $\{a, b\}$ and υ is a 100×1 column vector of the unobserved term υ_i.

To solve the system we can use the same "division" idea that we used for standard algebra. In matrix algebra, we do division by multiplying the inverse of the matrix $(\mathbf{X})$ by both sides of the equation. Actually, this is how we do division in normal algebra, it is just that we gloss over it and say we "moved" it from one side of the equation to the other.

The problem is that our matrix is not invertible. Only **full-rank** square matrices are invertible and our matrix is not even square.[9] The solution is to create a **generalized matrix inverse**.

We can make our matrix $(\mathbf{X})$ square by pre-multiplying it by its **transpose**.[10] The notation for this is $\mathbf{X}'$. Also remember that in matrix multiplication order matters! Pre-multiplying means placing the transposed matrix to the left of the original matrix.

$$\mathbf{X}'\mathbf{y} = \mathbf{X}'\mathbf{X}\beta + \mathbf{X}'\upsilon \qquad (1.11)$$

The matrix $\mathbf{X}'\mathbf{X}$ is a 2×2 matrix as it is a 2×100 matrix multiplied by a 100×2 matrix.[11]

To solve for the parameters of the model (β), we pre-multiply both side of Equation (1.11) by the matrix we created. Thus we have the following equation.

$$(\mathbf{X}'\mathbf{X})^{-1}\mathbf{X}'\mathbf{y} = (\mathbf{X}'\mathbf{X})^{-1}\mathbf{X}'\mathbf{X}\beta + (\mathbf{X}'\mathbf{X})^{-1}\mathbf{X}'\upsilon \qquad (1.12)$$

Simplifying and rearranging we have the following linear algebra derivation of the model.

$$\beta = (\mathbf{X}'\mathbf{X})^{-1}\mathbf{X}'\mathbf{y} - (\mathbf{X}'\mathbf{X})^{-1}\mathbf{X}'\upsilon \qquad (1.13)$$

From this we have the matrix algebra based estimate of our model.

$$\hat{\beta} = (\mathbf{X}'\mathbf{X})^{-1}\mathbf{X}'\mathbf{y} \qquad (1.14)$$

Remember that we never observe υ and so we do not know the second part of Equation (1.13).[12]

We didn't just drop the unobserved term, we averaged over it. If the assumptions of OLS hold, then summation $(\mathbf{X}'\mathbf{X})^{-1}\mathbf{X}'\upsilon$ will generally be close to zero. This issue is discussed more in Chapter 3 and Appendix A.

---

[9]The rank of the matrix refers to the number of linearly independent columns (or rows). A matrix is full-rank if all of its columns (rows) are linearly independent of each other.

[10]A transpose of a matrix is the same matrix with the columns and rows swapped.

[11]Remember that for matrix multiplication the "inside" numbers need to match.

[12]We traditionally use the "hat" notation to represent that $\hat{\beta}$ is the estimate of β.

1.3.6   Matrix Estimator of OLS in R

We can follow a similar procedure for inverting a matrix using **R**. The matrix of explanatory variables includes a first column of 1's which accounts for the intercept term.

```
> X <- cbind(1,x) # remember the column of 1's
```

Next we need to transpose the matrix. To illustrate a matrix transpose consider a matrix **A** which is 3×2 (3 rows and 2 columns) and has values 1 to 6.

```
> A <- matrix(c(1:6),nrow=3)
> # creates a 3 x 2 matrix.
> A

     [,1] [,2]
[1,]    1    4
[2,]    2    5
[3,]    3    6

> # See how R numbers elements of the matrix.
> t(A)   # transpose of matrix A

     [,1] [,2] [,3]
[1,]    1    2    3
[2,]    4    5    6

> t(A)%*%A  # matrix multiplication of the transpose by itself

     [,1] [,2]
[1,]   14   32
[2,]   32   77
```

In our problem the transpose multiplied by itself gives the following 2×2 matrix.

```
> t(X)%*%X

                    x
  100.00000 49.56876
x  49.56876 32.99536
```

In **R**, the matrix inverse can be found using the `solve()` function.

```
> solve(t(X)%*%X)  # solves for the matrix inverse.

                     x
   0.03916481 -0.05883708
x -0.05883708  0.11869792
```

The matrix algebra presented Equation (1.14) is pseudo-code for the operation in **R**.

```
> beta_hat <- solve(t(X)%*%X)%*%t(X)%*%y
> beta_hat

     [,1]
  2.181997
x 3.092764
```

Our estimates are $\hat{a} = 2.18$ and $\hat{b} = 3.09$. These values are not equal to the true values of $a = 2$ and $b = 3$, but they are fairly close. Try running the simulation again, but changing N to 1,000. Are the new estimates closer to the their true values? Why?[13]

We can check that we averaged over the unobserved term to get something close to 0.

```
> solve(t(X)%*%X)%*%t(X)%*%u

      [,1]
  0.1819969
x 0.0927640
```

1.4   Least Squares Method for OLS

An alternative algorithm is least squares. This section presents the algebra of least squares and compares the linear model (lm()) estimator to the matrix algebra version derived above.

1.4.1   Moment Estimation

Least squares requires that we assume the unobserved characteristic has a mean of 0. We say that the **first moment** of the unobserved characteristic is 0. A moment refers to the expectation of a random variable taken to some power. The first moment is the expectation of the variable taken to the power of 1. The second moment is the expectation of the variable taken to the power of 2, etc.

$$\mathbb{E}(v_i) = 0 \tag{1.15}$$

From Equation (1.1) we can rearrange and substitute into Equation (1.15).

$$\mathbb{E}(y_i - a - bx_i) = 0 \tag{1.16}$$

---

[13]In the simulation the unobserved term has a mean of 0. As mentioned above, the average of the unobserved term gets closer to zero when N is larger.

Equation (1.16) states that the expected difference, or mean difference, between y_i and the predicted value of y_i $(a + bx_i)$ is 0.

The **sample analog** of the left-hand side of Equation (1.16) is the following average.

$$\frac{1}{N} \sum_{i=1}^{N} (y_i - a - bx_i) \tag{1.17}$$

The sample equivalent of the mean is the average. This approach to estimation is called **analog estimation** (Manski, 1988). We can make this number as close to zero as possible by minimizing the sum of squares, that is, finding the "least squares."

1.4.2   Algebra of Least Squares

Again it helps to illustrate the algorithm by solving the simpler problem. Consider a version of the problem in which we are just trying to estimate the b parameter.

We want to find the b such that Equation (1.16) holds. Our sample analog is to find the $\hat{b}$ that makes Equation (1.17) as close to zero as possible.

$$\min_{\hat{b}} \quad \frac{1}{N} \sum_{i=1}^{N} (y_i - 2 - \hat{b}x_i)^2 \tag{1.18}$$

We can solve for $\hat{b}$ by finding the first order condition of the problem.[14] That is, we want the $\hat{b}$ such that the following equality holds.

$$\frac{1}{N} \sum_{i=1}^{N} -2x_i(y_i - 2 - \hat{b}x_i) = 0 \tag{1.19}$$

Note that we can divide both sides by -2, giving the following rearrangement.

$$\hat{b} = \frac{\frac{1}{N} \sum_{i=1}^{N} x_i y_i - 2\frac{1}{N} \sum_{i=1}^{N} x_i}{\frac{1}{N} \sum_{i=1}^{N} x_i x_i} \tag{1.20}$$

Our estimate of the relationship b is equal to a measure of the covariance between X and Y divided by the variance of X. This is the sense that we can think of our OLS estimate as a measure of the correlation between the independent variable (the X's) and the outcome variable (the Y).

1.4.3   Estimating Least Squares in R

We can program the least squares estimator in two ways. First, we can solve the problem presented in Equation (1.18). Second, we can use the solution to the first order condition presented in Equation (1.20).

---

[14]Under certain conditions the optimal value, minimum or maximum, can be determined by the first order condition. The first order condition can be found by taking the derivative and setting the equation to zero.

```
> optimize(function(b) sum((y - 2 - b*x)^2), c(-10,10))$minimum
```

[1] 3.366177

```
> # optimize() is used when there is one variable.
> # note that the function can be defined on the fly
> # $minimum presents one of the outcomes from optimize()
```

We can use `optimize()` to find a single optimal value of a function. We can use `function()` to create the sum of squared difference function, which is the first argument of `optimize()`. The procedure searches over the interval which is the second argument. Here it searches over the real line from -10 to 10 (`c(-10,10)`).[15] It looks for the value that minimizes the sum of the squared differences. The result is 3.37. Why do you think this is so far from the true value of 3?[16]

Alternatively, we can use the first order condition.

```
> (mean(x*y) - 2*mean(x))/mean(x*x)
```

[1] 3.366177

Solving out Equation (1.20) in **R** gives an estimate of 3.37. Why do these two approaches give identical answers?

1.4.4   The `lm()` Function

The standard method for estimating OLS in **R** is to use the `lm()` function. This function creates an **object** in **R**. This object keeps track of various useful things such as the vector of parameter estimates.[17]

```
> data1 <- as.data.frame(cbind(y,x))
> # creates a data.frame() object which will be used in the
> # next section.
> lm1 <- lm(y ~ x)   # lm creates a linear model object
> length(lm1)
```

[1] 12

```
> # reports the number of elements of the list object
> # names(lm1) # reports the names of the elements
```

We can compare the answers from the two algorithms, the `lm()` procedure and the matrix algebra approach.

---

[15]This is arbitrary, but I know the true values lie between these two values.

[16]See the discussion above for the previous estimators.

[17]An object in a programming language is data saved in various formats that can be accessed by the code. See Appendix B for discussion of programming in **R**.

```
> lm1$coefficients  # reports the coefficient estimates

(Intercept)          x
  2.181997    3.092764

> # $ can be used to call a particular element from the list.
> # lm1[1] reports the same thing.
> t(beta_hat) # results from the matrix algebra.


                     x
[1,] 2.181997 3.092764
```

Comparing the two different approaches in **R**, we can see they give identical estimates. This is not coincidental. The `solve()` function is based on a least squares procedure. It turns out there is no real computational difference between an "algebra" approach and a "least squares" approach in **R**.

1.5 Measuring Uncertainty

How do we give readers a sense of how accurate our estimate is? The previous section points out that the estimated parameter is not equal to the true value and may vary quite a bit from the true value. Of course, in the simulations we know exactly what the true value is. In real world econometric problems, we do not.

This section provides a brief introduction to some of the issues that arise when thinking about reporting the uncertainty around our estimate. Appendix A discusses these issues in more detail.

1.5.1 Data Simulations

Standard statistical theory assumes that the data we observe is one of many possible samples. Luckily, with a computer we can actually see what happens if we observe many possible samples. The simulation is run 1,000 times. In each case a sample of 100 is drawn using the same parameters as above. In each case, OLS is used to estimate the two parameters $\hat{a}$ and $\hat{b}$.

```
> set.seed(123456789)
> K <- 1000
> sim_res <- matrix(NA,K,2)
> # creates a 1000 x 2 matrix filled with NAs.
> # NA denotes a missing number in R.
> # I prefer using NA rather than an actual number like 0.
> # It is better to create the object to be filled
```

```
> # prior to running the loop.
> # This means that R has the object in memory.
> # This simple step makes loops in R run a lot faster.
> for (k in 1:K) {
+    # the "for loop" starts at 1 and moves to K
+    x <- runif(N)
+    u <- rnorm(N)
+    y <- a + b*x + u
+    sim_res[k,] <- lm(y ~ x)$coefficients
+    # inputs the coefficients from simulated data into the
+    # kth row of the matrix.
+    # print(k)
+    # remove the hash to keep track of the loop
+ }
> colnames(sim_res) <- c("Est. of a", "Est. of b")
> # labels the columns of the matrix.

> # install.packages("xtable")
> # or you can install in RStudio by going to
> # "Packages > Install > xtable" in the
> # bottom left window.
> require(xtable)
> # require loads an R package from memory
> # note that the package needs to be installed.
> # the xtable package produces fairly nice latex tables for a
> # variety of different objects.
> # summary produces a standard summary of the matrix.
> sum_tab <- summary(sim_res)
> rownames(sum_tab) <- NULL   # no row names.
> # NULL creates an empty object in R.
> print(xtable(sum_tab), floating=FALSE)
```

	Est. of a	Est. of b
1	Min. :1.389	Min. :1.754
2	1st Qu.:1.875	1st Qu.:2.758
3	Median :2.015	Median :2.979
4	Mean :2.013	Mean :2.983
5	3rd Qu.:2.156	3rd Qu.:3.206
6	Max. :2.772	Max. :4.466

TABLE 1.1
Summary of 1,000 simulations of OLS estimates of samples with 100 observations. The true values are $a = 2$ and $b = 3$.

Table 1.1 summarizes the results of the simulation. On average the estimated parameters are quite close to the true values, differing by 0.013 and 0.017 respectively. These results suggest the estimator is **unbiased**. An estimator is unbiased if it is equal to the true value on average. The OLS estimator is unbiased in this simulation because the unobserved term is drawn from a distribution with a mean of 0.

The table also shows that our estimates could differ substantially from the true value. As an exercise, what happens when the sample size is decreased or increased? Try $N = 10$ or $N = 5,000$.

1.5.2   Introduction to the Bootstrap

It would be great to present the analysis above for any estimate. Of course we cannot do that because we do not know the true values of the parameters. The Stanford statistician, Brad Efron, argues we can use the **analogy principle**. We don't know the true distribution but we do know its sample analog, which is the sample itself.

Efron's idea is called **bootstrapping**. And no, I have no idea what a bootstrap is. It refers to the English saying, "to pull one's self up by his bootstraps." In the context of statistics, it means to use the statistical sample itself to create an estimate of how accurate our estimate is. Efron's idea is to repeatedly draw pseudo-samples from the actual sample, randomly and with replacement, and then for each pseudo-sample re-estimate the model.

If our sample is pretty large then it is a pretty good representation of the true distribution. We can create a sample analog version of the thought exercise above. We can recreate an imaginary sample and re-estimate the model on that imaginary sample. If we do this a lot of times we get a distribution of pseudo-estimates. This distribution of pseudo-estimates provides us with information on how uncertain our original estimate is.

1.5.3   Bootstrap in R

We can create a bootstrap estimator in **R**. First, we create a simulated sample data set. The bootstrap code draws repeatedly from the simulated sample to create 1000 pseudo-samples. The code then creates summary statistics of the estimated results from each of the pseudo-samples.

```
> set.seed(123456789)
> K <- 1000
> bs_mat <- matrix(NA,K,2)
> for (k in 1:K) {
+    index_k <- round(runif(N, min=1, max=N))
+    # creates a pseudo-random sample.
+    # draws N elements uniformly between 1 and N.
+    # rounds all the elements to the nearest integer.
+    data_k <- data1[index_k,]
```

```
+    bs_mat[k,] <- lm(y ~ x, data=data_k)$coefficients
+    # print(k)
+ }
> tab_res <- matrix(NA,2,4)
> tab_res[,1] <- colMeans(bs_mat)
> # calculates the mean for each column of the matrix.
> # inputs into the first column of the results matrix.
> tab_res[,2] <- apply(bs_mat, 2, sd)
> # a method for having the function sd() to act on each
> # column of the matrix.  Dimension 2 is the columns.
> # sd() calculates the standard deviation.
> tab_res[,3] <- quantile(bs_mat[,1],c(0.025,0.975))
> # calculates quantiles of the column at 2.5% and 97.5%.
> tab_res[,4] <- quantile(bs_mat[,2],c(0.025,0.975))
> colnames(tab_res) <- c("Mean", "SD", "2.5%", "97.5%")
> rownames(tab_res) <- c("Est. of a","Est. of b")
> # labels the rows of the matrix.
```

	Mean	SD	2.5%	97.5%
Est. of a	2.21	0.21	1.79	2.35
Est. of b	3.06	0.36	2.65	3.74

TABLE 1.2
Bootstrapped estimates from the simulation.

Table 1.2 presents the bootstrapped estimates of the OLS model on the simulated data. The table presents the mean of the estimates from the pseudo-samples, the standard deviation of the estimates and the range which includes 95% of the cases. It is good to see that the true values do lie in the 95% range.

1.5.4   Standard Errors

```
> print(xtable(summary(lm1)), floating=FALSE)
```

	Estimate	Std. Error	t value	$\Pr(>|t|)$
(Intercept)	2.1820	0.1825	11.96	0.0000
x	3.0928	0.3177	9.74	0.0000

TABLE 1.3
OLS results from the simulation.

Table 1.3 presents the standard results that come out of the lm() procedure in **R**. Note that these are not the same as the bootstrap estimates.[18] The second column provides information on how uncertain the estimates are. It gives the standard deviation of the imaginary estimates assuming that the imaginary estimates are normally distributed. The last two columns present information which may be useful for **hypothesis testing**. This is discussed in detail in Appendix A.

The lm() procedure assumes the difference between the true value and the estimated value is distributed normally. That turns out to be the case in this simulation. In the real world there may be cases where it may be less reasonable to make this assumption.

1.6 Returns to Schooling

Now we have the basics of OLS, we can start on a real problem. One of the most studied areas of microeconometrics is returns to schooling (Card, 2001). Do policies that encourage people to get more education, improve their economic outcomes? One way to answer this question is to use survey data to determine how much education a person received and how much income they earned.

Berkeley labor economist David Card, analyzed this question using survey data on men aged between 14 and 24 in 1966 (Card, 1995). The data set provides information on each man's years of schooling and income. These are measured ten years later, in 1976.

1.6.1 A Linear Model of Returns to Schooling

Card posits that income in 1976 is determined by the individual's years of schooling.

$$\text{Income}_i = \alpha + \beta \text{Education}_i + \text{Unobserved}_i \qquad (1.21)$$

In Equation (1.21), income in 1976 for individual i is determined by their education level and other unobserved characteristics such as the unemployment rate in the place they live. We want to estimate β.

1.6.2 NLSM Data

The National Longitudinal Survey of Older and Younger Men (NLSM) is a survey data set used by David Card. Card uses the young cohort who were in the late teens and early twenties in 1966. Card's version can be downloaded from his website, http://davidcard.berkeley.edu/data_sets.

---

[18]The derivation of these numbers is discussed in Appendix A.

html.[19] My version of the data is found here: `https://sites.google.com/view/microeconometricswithr/table-of-contents`. Note before running this code you need to create a script, say "Card.R" and save that file with your data. You can then go to the `RStudio` menu and set the working directory to the location of your script (Session -> Set Working Directory -> To Source File Location). Alternatively, you can use the command `setwd()`.

```
> x <- read.csv("nls.csv",as.is=TRUE)
> # I follow the convention of defining any data set as x
> # (or possibly y or z).
> # I set the working directory to the one where my main
> # script is saved, which is the same place as the data.
> # Sessions -> Set Working Directory -> Source File Location
> # It is important to add "as.is = TRUE",
> # otherwise R may change your variables into "factors"
> # which is confusing and leads to errors.
> # Factors can be useful when you want to instantly create
> # a large number of dummy variables from a single variable.
> # However, it can make things confusing if you don't realize
> # what R is doing.
> x$wage76 <- as.numeric(x$wage76)
> x$lwage76 <- as.numeric(x$lwage76)
> # changes format from string to number.
> # "el wage 76" where "el" is for "log"
> # Logging helps make OLS work better.  This is because wages
> # have a skewed distribution, and log of wages do not.
> x1 <- x[is.na(x$lwage76)==0,] # creates a new data set
> # with missing values removed
> # is.na() determines missing value elements ("NA"s)
```

After reading in the data, the code changes the format of the data. Because some observations have missing values, the variables will import as strings. Changing the string to a number creates a missing value code `NA`. The code then drops observations with the missing variables.

1.6.3   Plotting Returns to Schooling

Figure 1.3 presents a simple plot of the relationship between log wages in 1976 and the years of schooling. Again we see a distinct positive relationship even though the observations are spread out in a cloud. We can see that people who do not graduate from high school (finish with less than 12 years of education) earn less on average than those who attend college (have more than 12 years of education). There is a lot of overlap between the distributions.

---

[19]This and other similar surveys can also be found at the US Bureau of Labor Statistics (`https://www.bls.gov/nls/home.htm`). Note however, this particular data set was not available when the site was accessed in October 2020.

```
> lm1 <- lm(lwage76 ~ ed76, data=x1)
> plot(x1$ed76,x1$lwage76, xlab="Years of Education",
+       ylab="Log Wages (1976)")
> # plot allows us to label the charts
> abline(a=lm1$coefficients[1],b=lm1$coefficients[2],lwd=3)
```

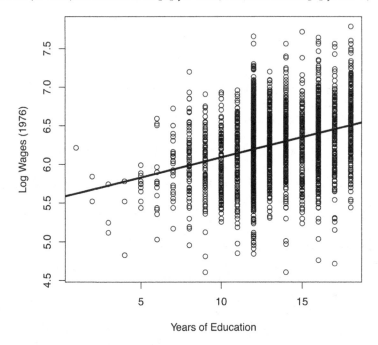

FIGURE 1.3
Plot of years of education and log wages in 1976 with the OLS estimate
represented by the line.

1.6.4   Estimating Returns to Schooling

	Estimate	Std. Error	t value	Pr($>$|t|)
(Intercept)	5.5709	0.0388	143.47	0.0000
ed76	0.0521	0.0029	18.15	0.0000

TABLE 1.4
OLS regression of log wages from 1976 on the number of years of schooling up
to 1976.

We can use OLS to estimate the average effect of schooling on income.

Table 1.4 gives the OLS estimate. The coefficient estimate of the relationship between years of schooling and log wages is 0.052. The coefficient is traditionally interpreted as the percentage increase in wages associated with a 1 year increase in schooling (Card, 1995). This isn't quite right, but it is pretty close. Below shows the predicted percentage change in wages, measured at the mean of wages, being 5.4%. Compare this to reading the coefficient as 5.2%.

```
> exp(log(mean(x1$wage76)) + lm1$coef[2])/mean(x1$wage76)

    ed76
1.053475
```

This estimate suggests high returns to schooling. However, the model makes a number of important assumptions about how the data is generated. The following chapters discuss the implications of those assumptions in detail.

1.7   Discussion and Further Reading

OLS is the "go to" method for estimation in microeconometrics. The method makes two strong assumptions. First, the unobserved characteristics must be **independent** of the policy variable. Second, the unobserved characteristics must affect the outcome variable **additively**. These two assumptions allow averaging to disentangle the causal effect from the effect of unobserved characteristics. We can implement the averaging using one of two algorithms, a matrix algebra-based algorithm or the least squares algorithm. In **R**, the two algorithms are computationally equivalent. The next two chapters consider weakening the assumptions. Chapter 2 takes the first step by allowing the OLS model to include more variables. Chapter 3 considers cases where neither assumption holds.

Measuring and reporting uncertainty has been the traditional focus of statistics and econometrics. This book takes a more "modern" approach to focus attention on issues around measuring and determining causal effects. This chapter takes a detour into the traditional areas. It introduces the bootstrap method and discusses traditional standard errors. These topics are discussed in more detail in the Appendix.

The chapter introduces causal graphs. Pearl and Mackenzie (2018) provide an excellent introduction to the power of this modeling approach.

To understand more about the returns to schooling literature, I recommend Card (2001). Chapters 2 and 3 replicate much of the analysis presented in Card (1995). The book returns to the question of measuring returns to schooling in Chapters 2, 3, 6, 8 and 11.

2

Multiple Regression

2.1 Introduction

In Chapter 1 there is only one **explanatory** variable. However, in many questions we expect multiple explanations. In determining a person's income, the education level is very important. There is clear evidence that other factors are also important, including experience, gender and race. Do we need to account for these factors in determining the effect of education on income?

Yes. In general, we do. Goldberger (1991) characterizes the problem as one of running a **short regression** when the data is properly explained by a **long regression**. This chapter discusses when we should and should not run a **long regression**. It also discusses an alternative approach. Imagine a policy that can affect the outcome either directly or indirectly through another variable. Can we estimate both the direct effect and the indirect effect? The chapter combines OLS with **directed acyclic graphs** (DAGs) to determine *how* a policy variable affects the outcome. It then illustrates the approach using actual data on mortgage lending to determine whether bankers are racist or greedy.

2.2 Long and Short Regression

Goldberger (1991) characterized the problem of omitting explanatory variables as the choice between long and short regression. The section considers the relative accuracy of long regression in two cases: when the explanatory variables are independent of each other, and when the explanatory variables are correlated.

2.2.1 Using Short Regression

Consider an example where true effect is given by the following long regression. The dependent variable y is determined by both x and w (and v).

$$y_i = a + bx_i + cw_i + v_i \tag{2.1}$$

In our running example, think of y_i as representing the income of individual i. Their income is determined by their years of schooling (x_i) and by their experience (w_i). It is also determined by unobserved characteristics of the individual (v_i). The last may include characteristics of where they live such as the current unemployment rate.

We are interested in estimating b. We are interested in estimating returns to schooling. In Chapter 1, we did this by estimating the following **short regression**.

$$y_i = a + bx_i + v_{wi} \qquad (2.2)$$

By doing this we have a different "unobserved characteristic." The unobserved characteristic, v_{wi} combines both the effects of the unobserved characteristics from the long regression (v_i) and the potentially observed characteristic (w_i). We know that experience affects income, but in Chapter 1 we ignore the possibility.

Does it matter? Does it matter if we just leave out important **explanatory** variables? Yes. And No. Maybe. It depends. What was the question?

2.2.2   Independent Explanatory Variables

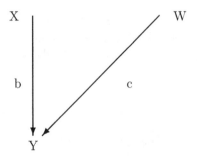

FIGURE 2.1
Two variable causal graph.

Figure 2.1 presents the independence case. There are two variables that determine the value of Y, these are X and W. If Y is income, then X may be schooling while W is experience. However, the two effects are independent of each other. In the figure, independence is represented by the fact that there is no line joining X to W, except through Y.

Consider the simulation below and the results of the various regressions presented in Table 2.1. Models (1) and (2) show that it makes little difference if we run the short or long regression. Neither of the estimates is that close, but that is mostly due to the small sample size. It does impact the estimate of the constant; can you guess why?

2.2.3 Dependent Explanatory Variables

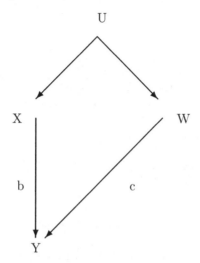

FIGURE 2.2
Two variable causal graph with dependence.

Short regressions are much less trustworthy when there is some sort of dependence between the two variables. Figure 2.2 shows the causal relationship when X and W are related to each other. In this case, a short regression will give a biased estimate of b because it will incorporate c. It will incorporate the effect of W. Pearl and Mackenzie (2018) call this a **backdoor relationship** because you can trace a relationship from X to Y through the backdoor of U and W.

Models (3) and (4) of Table 2.1 present the short and long estimators for the case where there is dependence. In this case we see a big difference between the two estimators. The long regression gives estimates of b and c that are quite close to the true values. The short regression estimate of b is considerably larger than the true value. In fact, you notice the **backdoor relationship**. The estimate of b is close to 7 which is the combined value of b and c.

2.2.4 Simulation with Multiple Explanatory Variables

The first simulation assumes that x and w affect y **independently**. That is, while x and w both affect y, they do not affect each other nor is there a common factor affecting them. In the simulations, the common factor is determined by the weight α. In this case, there is zero weight placed on the common factor potentially affecting both observed characteristics.

```
> set.seed(123456789)
> N <- 1000
> a <- 2
> b <- 3
> c <- 4
> u_x <- rnorm(N)
> alpha <- 0
> x <- x1 <- (1 - alpha)*runif(N) + alpha*u_x
> w <- w1 <- (1 - alpha)*runif(N) + alpha*u_x
> # this creates two identical variables
> # x1 and w1 are used later.
> u <- rnorm(N)
> y <- a + b*x + c*w + u
> lm1 <- lm(y ~ x)
> lm2 <- lm(y ~ x + w)
```

The second simulation allows for **dependence** between x and w. In the simulation this is captured by a positive value for the weight that each characteristic places on the common factor u_x. Table 2.1 shows that in this case it makes a big difference if a short or long regression is run. The short regression gives a biased estimate because it is accounting for the effect of w on y. Once this is accounted for in the long regression, the estimates are pretty close to the true value.

```
> alpha <- 0.5
> x <- x2 <- (1 - alpha)*runif(N) + alpha*u_x
> w <- w2 <- (1 - alpha)*runif(N) + alpha*u_x
> y <- a + b*x + c*w + u
> lm3 <- lm(y ~ x)
> lm4 <- lm(y ~ x + w)
```

The last simulation suggests that we need to take care not to overly rely on long regressions. If x and w are highly correlated, then the short regression gives the dual effect, while the long regression presents garbage.[1]

```
> alpha <- 0.95
> x <- x3 <- (1 - alpha)*runif(N) + alpha*u_x
> w <- w3 <- (1 - alpha)*runif(N) + alpha*u_x
> y <- a + b*x + c*w + u
> lm5 <- lm(y ~ x)
> lm6 <- lm(y ~ x + w)
```

---

[1] Actually, the combined effect is captured by adding the two coefficients together. The model can't separate the two effects.

```
> # install.packages("stargazer")
> require(stargazer)  # this package is very good at
> # presenting regression results.
> stargazer(list(lm1,lm2,lm3,lm4,lm5,lm6),
+          keep.stat = c("n","rsq"),
+          float = FALSE, font.size = "small", digits=2,
+          keep=c(1:6))
```

	Dependent variable:					
			y			
	(1)	(2)	(3)	(4)	(5)	(6)
x	3.14***	2.81***	6.84***	2.86***	7.01***	0.67
	(0.17)	(0.11)	(0.08)	(0.17)	(0.03)	(1.59)
w		4.05***		4.16***		6.35***
		(0.11)		(0.16)		(1.59)
Constant	3.98***	2.15***	2.14***	2.07***	2.07***	2.08***
	(0.10)	(0.08)	(0.05)	(0.04)	(0.03)	(0.03)
Observations	1,000	1,000	1,000	1,000	1,000	1,000
R^2	0.26	0.68	0.88	0.93	0.98	0.98
Note:					*p<0.1; **p<0.05; ***p<0.01	

```
> # to see this in the console set type = "text"
```

TABLE 2.1
OLS estimates of short and long regressions. Models (1) and (2) compare short and long regressions when there is no dependence between x and w. Models (3) and (4) present the same comparison with dependence. Models (5) and (6) are for the strong dependence case. In order, from top to bottom, the true values are 3, 4 and 2.

Table 2.1 shows what happens when you run a short regression with dependence between the variables.[2] When there is no dependence the short regression does fine, actually a little better in this example. However, when there is dependence the short regression is capturing both the effect of x and the effect of w on y. Running long regressions is not magic. If the two variables are strongly correlated then the long regression cannot distinguish between the two different effects. There is a **multicollinearity** problem. This issue is discussed in more detail below.

2.2.5   Matrix Algebra of Short Regression

To illustrate the potential problem with running a short regression consider the matrix algebra.

$$\mathbf{y} = \mathbf{X}\beta + \mathbf{W}\gamma + \upsilon \tag{2.3}$$

Equation (2.3) gives the true relationship between the outcome vector y and the observed explanatory variables $\mathbf{X}$ and $\mathbf{W}$. "Dividing" by $\mathbf{X}$ we have the following difference between the true short regression and the estimated short regression.

$$\hat{\beta} - \beta = (\mathbf{X}'\mathbf{X})^{-1}\mathbf{X}'\mathbf{W}\gamma + (\mathbf{X}'\mathbf{X})^{-1}\mathbf{X}'\upsilon \tag{2.4}$$

Equation (2.4) shows that the short regression gives the same answer if $(\mathbf{X}'\mathbf{X})^{-1}\mathbf{X}'\upsilon = 0$ and either $\gamma = 0$ or $(\mathbf{X}'\mathbf{X})^{-1}\mathbf{X}'\mathbf{W} = 0$. The first will tend to be close to zero if the Xs are independent of the unobserved characteristic. This is the standard assumption for ordinary least squares. The second will tend to be close to zero if either of the W's have no effect on the outcome (Y) or there is no correlation between the Xs and the Ws. The linear algebra illustrates that the extent that the short regression is biased depends on both the size of the parameter γ and the correlation between X and W. The correlation is captured by $\mathbf{X}'\mathbf{W}$.

```
> cov(x1,w1)  # calculates the covariance between x1 and w1

[1] 0.007019082

> cov(x2,w2)

[1] 0.2557656

> t(x2)%*%w2

          [,1]
[1,] 318.8261

> # this corresponds to the linear algebra above
> # it measures the correlation between the Xs and Ws.
```

---

[2]The table uses the stargazer package (Hlavac, 2018). If you are using stargazer in Sweave then start the chunk with results=tex embedded in the chunk header.

In our simulations we see that in the first case the covariance between x and w is small, while for the second case it is much larger. It is this covariance that causes the short regression to be biased.

2.3 Collinearity and Multicollinearity

As we saw in Chapter 1, the true parameter vector can be written as follows.

$$\beta = (\mathbf{X}'\mathbf{X})^{-1}\mathbf{X}'\mathbf{y} - (\mathbf{X}'\mathbf{X})^{-1}\mathbf{X}'\upsilon \qquad (2.5)$$

where here $\mathbf{X}$ is a matrix that includes columns for x and w, while the parameter vector (β) includes the effect of both x and w. The estimated $\hat{\beta}$ is the first part and the difference between the true value and the estimated value is the second part.

Chapter 1 states that in order to interpret the parameter estimates as measuring the true effects of x and y, we need two assumptions. The first assumption is that the unobserved characteristic is independent of the observed characteristic. The second assumption is that the unobserved characteristic affects the dependent variable additively. However, these two assumptions are not enough. In order for this model to provide reasonable estimates, the matrix $\mathbf{X}'\mathbf{X}$ must be invertible. That is, our matrix of observable characteristics must be **full-column rank**.[3]

In statistics, the problem that the our matrix of observables is not full-column rank is called "collinearity." Two, or more, columns are "co-linear." Determining whether a matrix is full-column rank is not difficult. If the matrix $\mathbf{X}$ is not full-column rank, then the program will throw an error and it will not compute. A more insidious problem is where the matrix of observed characteristics is "almost" not full-column rank. This problem is called **multicollinearity**.

Econometrics textbooks are generally not a lot of laughs. One prominent exception is Art Goldberger's *A Course in Econometrics* and its chapter on Multicollinearity. Goldberger points out that multicollinearity has many syllables but in the end it is just a problem of not having enough data (Goldberger, 1991). More accurately, a problem of not having enough variation in the data. He then proceeds by discussing the analogous problem of **micronumerosity**.[4] Notwithstanding Goldberger's jokes, multicollinearity is no joke.

Models (5) and (6) in Table 2.1 show what happens when the explanatory variables x and w are "too" dependent. The long regression seems to present garbage results. The short regression still gives biased estimates incorporating both the effect of x and w. But the long regression cannot disentangle the two effects.

---

[3]Matrix rank refers to the number of linearly independent columns or rows.
[4]Micronumerosity refers to the problem of small sample size.

2.3.1   Matrix Algebra of Multicollinearity

In the problem we make two standard assumptions. First, the average value
of the unobserved characteristic is 0. Second, the Xs are independent of the
Us. In matrix algebra this second assumption implies that $\mathbf{X}'v$ will be zero
when the sample size is large. Because of these two assumptions, the weighted
averages will tend to 0. However, when the matrix of observable characteristics
is "almost" not full-column rank then this weighted average can diverge quite
a lot from 0.

2.3.2   Understanding Multicollinearity with R

Given the magic of our simulated data, we can look into what is causing the
problem with our estimates.

```
> X2 <- cbind(1,x3,w3)
> solve(t(X2)%*%X2)%*%t(X2)%*%u

          [,1]
     0.0766164
x3  -2.3321265
w3   2.3462160
```

First, we can look at the difference between β and $\hat{\beta}$. In a perfect world, this
would be a vector of 0s. Here it is not close to that.

```
> mean(u)

[1] 0.07635957

> cov(x3,u)   # calculates the covariace between two variables

[1] 0.01316642

> cov(w3,u)

[1] 0.01413717

> (1/N)*t(X2)%*%u

          [,1]
     0.07635957
x3  0.01593319
w3  0.01687792
```

Again we can look at the main OLS assumptions, that the mean of the
unobserved term is zero and the covariance between the unobserved term and
the observed terms is small.

The operation above shows that the average of the unobserved characteristic does differ somewhat from 0. Still, it is not large enough to explain the problem. We can also look at the independence assumption, which implies that the covariance between the observed terms and the unobserved term will be zero (for large samples). Here, they are not quite zero, but still small. Again, not enough to explain the huge difference.

The problem is that we are dividing by a very small number. The inverse of a 2×2 matrix is determined by the following formula.

$$\begin{bmatrix} a & b \\ c & d \end{bmatrix}^{-1} = \frac{1}{ad - bc} \begin{bmatrix} d & -b \\ -c & a \end{bmatrix} \qquad (2.6)$$

where the rearranged matrix is divided by the determinant of the matrix. We see above that the reciprocal of the determinant of the matrix is very small. In this case, dividing by the very "small" inverse overwhelms everything else. This is the matrix equivalent of dividing by zero.

```
> 1/det(t(X2)%*%X2)
```

```
[1] 2.610265e-06
```

```
> # calculates the reciprocal of the determinant of the matrix.
```

2.4 Returns to Schooling

Now that we have a better idea of the value and risk of multiple regression, we can return to the question of returns to schooling. Card (1995) posits that income in 1976 is determined by a number of factors including schooling.

2.4.1 Multiple Regression of Returns to Schooling

We are interested in the effect of schooling on income. However, we want to account for how other variables may also affect income. Standard characteristics that are known to determine income are work experience, race, the region of the country where the individual grew up and the region where the individual currently lives.

$$\text{Income}_i = \alpha + \beta \text{Education}_i + \gamma \text{Experience}_i + \ldots + \text{Unobserved}_i \qquad (2.7)$$

In Equation (2.7), income in 1976 for individual i is determined by their education level, their experience, and other characteristics such as race, where the individual grew up and where the individual is currently living. We are interested in estimating β. In Chapter 1, we used a short regression to estimate $\hat{\beta} = 0.052$. What happens if we estimate a long regression?

2.4.2 NLSM Data

```
> x <- read.csv("nls.csv",as.is=TRUE)
> x$wage76 <- as.numeric(x$wage76)
> x$lwage76 <- as.numeric(x$lwage76)
> # note that this is "el" wage 76.
> x1 <- x[is.na(x$lwage76)==0,]
> x1$exp <- x1$age76 - x1$ed76 - 6 # working years after school
> x1$exp2 <- (x1$exp^2)/100
> # experienced squared divided by 100
> # this makes the presentation nicer.
```

The chapter uses the same data as Chapter 1. This time we create measures of experience. Each individual is assumed to have "potential" work experience equal to their age less years in full-time education. The code also creates a squared term for experience. This allows the estimator to capture the fact that wages tend to increase with experience but at a decreasing rate.

2.4.3 OLS Estimates of Returns to Schooling

```
> lm1 <- lm(lwage76 ~ ed76, data=x1)
> lm2 <- lm(lwage76 ~ ed76 + exp + exp2, data=x1)
> lm3 <- lm(lwage76 ~ ed76 + exp + exp2 + black + reg76r,
+           data=x1)
> lm4 <- lm(lwage76 ~ ed76 + exp + exp2 + black + reg76r +
+             smsa76r + smsa66r + reg662 + reg663 + reg664 +
+             reg665 + reg666 + reg667 + reg668 + reg669,
+           data=x1)
> # reg76 refers to living in the south in 1976
> # smsa refers to whether they are urban or rural in 1976.
> # reg refers to region of the US - North, South, West etc.
> # 66 refers to 1966.
```

Table 2.2 gives the OLS estimates of the returns to schooling. The estimates on the coefficient of years of schooling on log wages vary from 0.052 to 0.093. The table presents the results in a traditional way. It presents models with longer and longer regressions. This presentation style gives the reader a sense of how much the estimates vary with the exact specification of the model. Model (4) in Table 2.2 replicates Model (2) from Table 2 of Card (1995).[5]

The longer regressions suggest that the effect of education on income is larger than for the shortest regression, although the relationship is not increasing in the number of explanatory variables. The effect seems to stabilize at around 0.075.

Under the standard assumptions of the OLS model, we estimate that an

---

[5] As an exercise try to replicate the whole table. Note that you will need to carefully read the discussion of how the various variables are created.

	Dependent variable:			
	lwage76			
	(1)	(2)	(3)	(4)
ed76	0.052***	0.093***	0.078***	0.075***
	(0.003)	(0.004)	(0.004)	(0.003)
exp		0.090***	0.085***	0.085***
		(0.007)	(0.007)	(0.007)
exp2		−0.249***	−0.234***	−0.229***
		(0.034)	(0.032)	(0.032)
black			−0.178***	−0.199***
			(0.018)	(0.018)
reg76r			−0.150***	−0.148***
			(0.015)	(0.026)
Observations	3,010	3,010	3,010	3,010
R^2	0.099	0.196	0.265	0.300
Note:			$^*p<0.1$; $^{**}p<0.05$; $^{***}p<0.01$	

TABLE 2.2
OLS estimates of returns to schooling. The table does not include the coefficient estimates for the regional dummies which are included in model (4).

additional year of schooling **causes** a person's income to increase (by about) 7.5%.[6] Are the assumptions of OLS reasonable? Do you think that unobserved characteristics of the individual do not affect both their decision to attend college and their income? Do you think family connections matter for both of these decisions? The next chapter takes on these questions.

2.5 Causal Pathways

Long regressions are not always the answer. The next two sections present cases where people rely on long regressions even though the long regressions can lead them astray. This section suggests using **directed acyclic graphs** (DAGs) as an alternative to long regression.[7]

Consider the case against Harvard University for discrimination against Asian-Americans in undergraduate admissions. The Duke labor economist, Peter Arcidiacono, shows that Asian-Americans have a lower probability of being admitted to Harvard than White applicants.[8] Let's assume that the effect of race on Harvard admissions is causal. The questions is then how does this causal relationship work? What is the causal pathway? One possibility is that there is a direct causal relationship between race and admissions. That is, Harvard admissions staff use the race of the applicant when deciding whether to make them an offer. The second possibility is that the causal relationship is indirect. Race affects admissions, but indirectly. Race is **mediated** by some other observed characteristics of the applicants such as their SAT scores, grades or extracurricular activities.

This is not some academic question. If the causal effect of race on admissions is direct, then Harvard may be legally liable. If the causal effect of race on admissions is indirect, then Harvard may not be legally liable.

Arcidiacono uses long regression in an attempt to show Harvard is discriminating. This section suggests his approach is problematic. The section presents an alternative way to disentangle the direct causal effect from the indirect causal effect.

2.5.1 Dual Path Model

Figure 2.3 illustrates the problem. The figure shows there exist two distinct causal pathways for X on Y. There is a direct causal effect of X on Y which

---

[6]See discussion in Chapter 1 regarding interpreting the coefficient.

[7]The UCLA computer scientist, Judea Pearl, is a proponent of using DAGs in econometrics. These diagrams are models of how the data is generated. The associated algebra helps the econometrician and the reader determine the causal relationships and whether or not they can be estimated (Pearl and Mackenzie, 2018).

[8]https://studentsforfairadmissions.org/

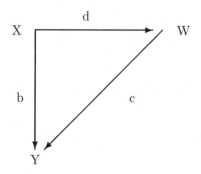

FIGURE 2.3
Dual path causal graph.

has value b. There is an indirect causal effect of X on Y which is mediated by W. The indirect effect of X on Y is c times d. The goal is to estimate all three parameters. Of particular interest is determining whether or not $b = 0$. That is, determine whether there exists is a direct effect of X on Y.

In algebra, we have the following relationship between X and Y.

$$y_i = a + bx_i + cw_i + v_i \tag{2.8}$$

and

$$w_i = dx_i + v_{wi} \tag{2.9}$$

Substituting Equation (2.9) into Equation (2.8) we have the full effect of X on Y.

$$y_i = a + (b + cd)x_i + v_i + cv_{wi} \tag{2.10}$$

The full relationship of X on Y $(b + cd)$ is straightforward to estimate. This means that if we can estimate c and d, then we can back out b.

Given the model described in Figure 2.3, it is straightforward to estimate $b + cd$ and it is straightforward to estimate d. It is not straightforward to estimate c. The issue is that there may be a **backdoor relationship** between W and Y. Running the regression of Y on W gives $c + \frac{b}{d}$.[9] It gives the direct relationship c plus the backdoor relationship $\frac{b}{d}$. The observed correlation between W and Y is determined by variation in X. OLS estimates c using the observed variation between W and Y but to some extent that variation is being determined by b and d. There are a number of solutions to this problem, including using the IV method discussed in the next chapter. Here, the problem is simplified.

---

[9]It is the reciprocal of d because we follow the arrow backwards from W to X.

For the remainder of the chapter we will make the problem go away by assuming that $b = 0$. Given this assumption, there is no **backdoor relationship** because X cannot affect the correlation between W and Y.[10] In the analysis below we can test the hypothesis that $b = 0$ under the maintained assumption that $b = 0$.[11]

2.5.2   Simulation of Dual Path Model

Consider the simulated data generated below. The data is generated according to the causal diagram in Figure 2.3.

```
> set.seed(123456789)
> N <- 50
> a <- 1
> b <- 0
> c <- 3
> d <- 4
> x <- round(runif(N)) # creates a vector of 0s and 1s
> u_w <- runif(N)
> w <- d*x + u_w
> u <- rnorm(N)
> y <- a + b*x + c*w + u
```

Table 2.3 presents the OLS results for the regression of Y on X. The results suggest that X has a very large effect on Y. However, look closely at the data that was generated. This effect is completely indirect. In the standard terminology, our regression suffers from **omitted variable bias**. The omitted variable is W.

	Estimate	Std. Error	t value	$\Pr(>|t|)$
(Intercept)	2.5867	0.2977	8.69	0.0000
x	12.2406	0.4129	29.65	0.0000

TABLE 2.3
OLS estimates of regressing y on x from the simulated data.

A standard solution to the **omitted variable problem** is to include the omitted variable in the regression (Goldberger, 1991). Table 2.4 presents results from a standard long regression. Remember that the true value of the coefficient on x is 0. Not only is the coefficient not equal to zero, a standard hypothesis test would reject that it is in fact equal to zero.

---

[10] You should think about the reasonableness of this assumption for the problem discussed below.
[11] See discussion of hypothesis testing in Appendix A.

	Estimate	Std. Error	t value	Pr($>$|t|)
(Intercept)	0.5644	0.2812	2.01	0.0505
x	-5.1481	1.8847	-2.73	0.0088
w	4.2854	0.4604	9.31	0.0000

TABLE 2.4
Results from the standard approach of a **long regression**.

The issue with the standard long regression is **multicollinearity**. From Figure 2.3 we see that x and w are causally related and thus they are correlated. That is "causation does imply correlation." This correlation between the independent variables makes it hard for the algorithm to separate the effect of x on y from the effect of w on y.

There is a better way to do the estimation. Figure 2.3 shows that there is a causal pathway from X to W. We can estimate d using OLS. In the simulation, the coefficient from that regression is close to 4. Note that the true value is 4. Similarly, the figure shows that there is a causal pathway from W to Y, labeled c. The true value of the effect is 3.

```
> e_hat <- lm(y ~ x)$coef[2]
> # element 2 is the slope coefficient of interest.
> c_hat <- lm(y ~ w)$coef[2]
> d_hat <- lm(w ~ x)$coef[2]
> # Estimate of b
> e_hat - c_hat*d_hat

        x
-0.08876039
```

By running three regressions, we can estimate the true value of b. First, running the regression of y on x we get $\hat{e} = \hat{b} + \hat{c}\hat{d}$. Second, we can run the regression of y on w to estimate $\hat{c}$. Third, we can run the regression of w on x to estimate $\hat{d}$. With these, we can back out $\hat{b}$. That is $\hat{b} = \hat{e} - \hat{c}\hat{d}$. Our new estimate of $\hat{b}$ will tend to be much closer to the true value of zero than the standard estimate from the long regression. Here it is -0.09 versus -5.15.

2.5.3 Dual Path Estimator Versus Long Regression

```
> set.seed(123456789)
> b_mat <- matrix(NA,100,3)
> for (i in 1:100) {
+    x <- round(runif(N))
+    u_w <- runif(N)
+    w <- d*x + u_w
```

```
+    u <- rnorm(N)
+    y <- a + b*x + c*w + u
+    lm2_temp <- summary(lm(y ~ x + w))
+    # summary provides more useful information about the object
+    # the coefficients object (item 4) provides additional
+    # information about the coefficient estimates.
+    b_mat[i,1] <- lm2_temp[[4]][2]
+    # The 4th item in the list is the results vector.
+    # The second item in that vector is the coefficient on x.
+    b_mat[i,2] <- lm2_temp[[4]][8]
+    # the 8th item is the T-stat on the coefficient on x.
+    e_hat <- lm(y ~ x)$coef[2]
+    c_hat <- lm(y ~ w)$coef[2]
+    d_hat <- lm(w ~ x)$coef[2]
+    b_mat[i,3] <- e_hat - c_hat*d_hat
+    # print(i)
+ }
> colnames(b_mat) <-
+    c("Standard Est","T-Stat of Standard","Proposed Est")

> summ_tab <- summary(b_mat)
> rownames(summ_tab) <- NULL
> print(xtable(summ_tab), floating=FALSE)
```

	Standard Est	T-Stat of Standard	Proposed Est
1	Min. :-5.1481	Min. :-3.02921	Min. :-0.119608
2	1st Qu.:-1.6962	1st Qu.:-0.80145	1st Qu.:-0.031123
3	Median :-0.3789	Median :-0.19538	Median :-0.008578
4	Mean :-0.1291	Mean :-0.07418	Mean :-0.002654
5	3rd Qu.: 1.3894	3rd Qu.: 0.74051	3rd Qu.: 0.030679
6	Max. : 6.5594	Max. : 2.68061	Max. : 0.104119

TABLE 2.5
Results comparing the standard method to the proposed method for 100 simulations. The table shows the estimates and the t-statistics for the standard method, as well as the estimates for the proposed method.

We can use simulation to compare estimators. We rerun the simulation 100 times and summarize the results of the two approaches in Table 2.5. The table shows that the proposed estimator gives values much closer to 0 than the standard estimator. It also shows that the standard estimator can, at times, provide misleading results. The estimate may suggest that the value of b is not only not 0, but statistically significantly different from 0. In contrast, the proposed estimate provides values that are always close to 0.

The proposed estimator is much more accurate than the standard estimator. Figure 2.4 shows the large difference in the accuracy of the two estimates. The standard estimates vary from over -5 to 5, while the proposed estimates have a much much smaller variance. Remember the two approaches are estimated on exactly the same data.

```
> hist(b_mat[,1],freq=FALSE,xlab="Estimate of b",main="")
> # plots the histogram, xlab labels the x-axis,
> # freq type of histogram.
> # main provide a title for the chart, here it is empty.
> lines(density(b_mat[,1]), type="l",lwd=3)
> # lines is used to add another plot to an existing plot.
> # calculates an approximation of the density function.
> # type l means that the plot is a line
> # lwd determines the line width.
> abline(v=c(min(b_mat[,3]),max(b_mat[,3])), lty=2, lwd=3)
> # v determines vertical lines.
```

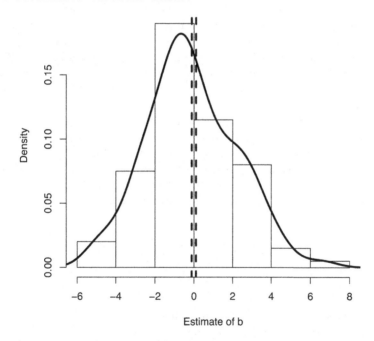

FIGURE 2.4

Histogram and density plot of standard estimator compared to the minimum and maximum values from the proposed estimator (dashed lines) from simulated data.

2.5.4   Matrix Algebra of the Dual Path Estimator

We can write out the dual path causal model more generally with matrix algebra.

$$
\begin{aligned}
\mathbf{y} &= \mathbf{X}\beta + \mathbf{W}\gamma + \epsilon \\
\mathbf{W} &= \mathbf{X}\Delta + \mathbf{\Upsilon}
\end{aligned} \tag{2.11}
$$

where $\mathbf{y}$ is a $N \times 1$ vector of the outcomes of interest, $\mathbf{X}$ is a $N \times J$ matrix representing the different variables that may be directly or indirectly determining the outcome, $\mathbf{W}$ is a $N \times K$ matrix representing the variables that are **mediating** the causal effect of the Xs on Y. The parameters of interest are β representing the direct effect and both γ and Δ representing the indirect effect. Note that while β and γ are $J \times 1$ and $K \times 1$ vectors respectively, Δ is a $J \times K$ matrix representing all the different effects of $\mathbf{X}$ on $\mathbf{W}$. Similarly ϵ is the $N \times 1$ vector of unobserved characteristics affecting the outcome and $\mathbf{\Upsilon}$ is a $N \times K$ matrix of all the unobserved characteristics determining the different elements of $\mathbf{W}$.

Equation (2.11) presents the model of the data generating process. In the simultaneous equation system we see that the matrix $\mathbf{X}$ enters both directly into the first equation and indirectly through the second equation. Note that the second equation is actually a set of equations representing all the different effects that the values represented by X have on the values represented by W. In addition, the unobserved characteristics that affect $\mathbf{y}$ are independent of unobserved characteristics that affect $\mathbf{W}$.

We are interested in estimating β, but we have learned above that the short regression of Y on X may give the wrong answer and the long regression of Y on X and W will also give the wrong answer. The solution is to estimate $\tilde{\beta}$ which represents the vector of coefficients from the short regression of Y on X. Then remember that we can derive β by estimating Δ and γ, as $\beta = \tilde{\beta} - \Delta\gamma$.

All three vectors can be estimated following the same matrix algebra we presented for estimating OLS.

$$
\begin{aligned}
\hat{\tilde{\beta}} &= (\mathbf{X}'\mathbf{X})^{-1}\mathbf{X}'\mathbf{y} \\
\hat{\Delta} &= (\mathbf{X}'\mathbf{X})^{-1}\mathbf{X}'\mathbf{W} \\
\hat{\gamma} &= (\mathbf{W}'\mathbf{W})^{-1}\mathbf{W}'\mathbf{y}
\end{aligned} \tag{2.12}
$$

The first equation $\hat{\tilde{\beta}}$ represents the "total effect" of the Xs on Y. It is the short regression of Y on X. The second equation follows from "dividing" by $\mathbf{X}$ in the second line of Equation (2.11). The third equation is the short regression of Y on W.

Substituting the results of the last two regressions into the appropriate places we get our proposed estimator for the direct effect of X on Y.

$$
\hat{\beta} = (\mathbf{X}'\mathbf{X})^{-1}\mathbf{X}'\mathbf{y} - (\mathbf{X}'\mathbf{X})^{-1}\mathbf{X}'\mathbf{W}(\mathbf{W}'\mathbf{W})^{-1}\mathbf{W}'\mathbf{y} \tag{2.13}
$$

Equation (2.13) presents the estimator of the direct effect of the Xs on Y.

Note that our proposed estimate of γ may be biased if the direct effect of X on Y is non-zero.[12]

2.5.5 Dual Path Estimator in R

Equation (2.13) is pseudo-code for the dual-path estimator in **R**.

```
> X <- cbind(1,x)
> W <- cbind(1,w)
> beta_tilde_hat <- solve(t(X)%*%X)%*%t(X)%*%y
> Delta_hat <- solve(t(X)%*%X)%*%t(X)%*%W
> gamma_hat <- solve(t(W)%*%W)%*%t(W)%*%y
> beta_tilde_hat - Delta_hat%*%gamma_hat

          [,1]
  -0.01515376
x  0.02444155
```

The estimated value of $\hat{\beta} = \{-0.015, 0.024\}$ is pretty close to the true value of $\beta = \{0, 0\}$.

2.6 Are Bankers Racist or Greedy?

Blacks are substantially more likely to be denied mortgages than Whites. Considering data used by Munnell et al. (1996), being black is associated with a 20% reduction in the likelihood of getting a mortgage.[13] The US Consumer Financial Protection Bureau states that the Fair Housing Act may make it illegal to refuse credit based on race.[14]

Despite these observed discrepancies, lenders may not be doing anything illegal. It may not be illegal to deny mortgages based on income or credit history. Bankers are allowed to maximize profits. They are allowed to deny mortgages to people that they believe are at high risk of defaulting. There may be observed characteristics of the applicants that are associated with a high risk of defaulting that are also associated with race.

Determining the causal pathway has implications for policy. If the effect of race is direct, then laws like the Fair Housing Act may be the correct policy response. If the effect is indirect, then such a policy will have little effect on a policy goal of increasing housing ownership among African Americans.

The section finds that there may be no direct effect of race on mortgage lending.

---

[12]See earlier discussion.

[13]The code book for the data set is located here: https://sites.google.com/view/microeconometricswithr/table-of-contents

[14]https://www.consumerfinance.gov/fair-lending/

2.6.1   Boston HMDA Data

The data used here comes from Munnell et al. (1996). This version is down-loaded from the data sets for Stock and Watson (2011) here: https://wps.pearsoned.com/aw_stock_ie_3/178/45691/11696965.cw/index.html. You can also download a csv version of the data from https://sites.google.com/view/microeconometricswithr/table-of-contents.

```
> x <- read.csv("hmda_aer.csv", as.is = TRUE)
> x$deny <- ifelse(x$s7==3,1,NA)
> # ifelse considers the truth of the first element.  If it is
> # true then it does the next element, if it is false it does
> # the final element.
> # You should be careful and make sure that you don't
> # accidentally misclassify an observation.
> # For example, classifying "NA" as 0.
> # Note that == is used in logical statments for equal to.
> x$deny <- ifelse(x$s7==1 | x$s7==2,0,x$deny)
> # In logical statements | means "or" and & means "and".
> # The variable names refer to survey questions.
> # See codebook at
> # https://sites.google.com/view/microeconometricswithr
> x$black <- x$s13==3 # we can also create a dummy by using
> # true/false statement.
```

The following table presents the raw effect of race on mortgage denials from the Munnell et al. (1996) data. It shows that being Black reduces the likelihood of a mortgage by 20 percentage points.

	Estimate	Std. Error	t value	Pr($>$|t|)
(Intercept)	0.1131	0.0069	16.30	0.0000
blackTRUE	0.1945	0.0174	11.21	0.0000

TABLE 2.6
OLS results of mortgage denials on race using data from Munnell et al. (1996).

2.6.2   Causal Pathways of Discrimination

Race may be affecting the probability of getting a mortgage through two different causal pathways. There may be a direct effect in which the lender is denying a mortgage because of the applicant's race. There may be an indirect effect in which the lender denies a mortgage based on factors such as income or credit history. Race is associated with lower income and poorer credit histories.

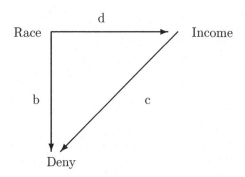

FIGURE 2.5
Dual path causal graph of mortgage denials.

Figure 2.5 represents the estimation problem. The regression in Table 2.6 may be picking up both the direct effect of Race on Deny (*b*) and the indirect effect of Race on Deny mediated by Income (*dc*).

2.6.3 Estimating the Direct Effect

```
> x$lwage <- NA
> x[x$s31a > 0 & x$s31a < 999999,]$lwage <-
+   log(x[x$s31a > 0 & x$s31a < 999999,]$s31a)
> # another way to make sure that NAs are not misclassified.
> # See the codebook for missing data codes.
> x$mhist <- x$s42
> x$chist <- x$s43
> x$phist <- x$s44
> x$emp <- x$s25a
> x$emp <- ifelse(x$emp > 1000, NA, x$emp)
```

To determine the causal effect of race we can create a number of variables from the data set that may mediate race. These variables measure information about the applicant's income, employment history and credit history.

```
> Y1 <- x$deny
> X1 <- cbind(1,x$black)
> W1 <- cbind(1,x$lwage,x$chist,x$mhist,x$phist,x$emp)
> index <- is.na(rowSums(cbind(Y1,X1,W1)))==0
> # this removes missing values.
> X2 <- X1[index,]
> Y2 <- Y1[index]
```

```
> W2 <- W1[index,]
> beta_tilde_hat <- solve(t(X2)%*%X2)%*%t(X2)%*%Y2
> Delta_hat <- solve(t(X2)%*%X2)%*%t(X2)%*%W2
> gamma_hat <- solve(t(W2)%*%W2)%*%t(W2)%*%Y2
> beta_tilde_hat - Delta_hat%*%gamma_hat

            [,1]
[1,] -0.01797654
[2,]  0.11112066
```

Adding these variables reduces the possible direct effect of race on mortgage denials by almost half. Previously, the analysis suggested that being African American reduced the probability of getting a mortgage by 20 percentage points. This analysis shows that at least 8 percentage points of that is due to an indirect causal effect mediated by income, employment history, and credit history. Can adding in more such variables reduce the estimated direct causal effect of race to zero?

2.6.4 Adding in More Variables

```
> x$married <- x$s23a=="M"
> x$dr <- ifelse(x$s45>999999,NA,x$s45)
> x$clines <- ifelse(x$s41>999999,NA,x$s41)
> x$male <- x$s15==1
> x$suff <- ifelse(x$s11 > 999999, NA, x$s11)
> x$assets <- ifelse(x$s35 > 999999, NA, x$s35)
> x$s6 <- ifelse(x$s6 > 999999, NA, x$s6)
> x$s50 <- ifelse(x$s50 > 999999, NA, x$s50)
> x$s33 <- ifelse(x$s33 > 999999, NA, x$s33)
> x$lr <- x$s6/x$s50
> x$pr <- x$s33/x$s50
> x$coap <- x$s16==4
> x$school <- ifelse(x$school > 999999, NA, x$school)
> x$s57 <- ifelse(x$s57 > 999999, NA, x$s57)
> x$s48 <- ifelse(x$s48 > 999999, NA, x$s48)
> x$s39 <- ifelse(x$s39 > 999999, NA, x$s39)
> x$chval <- ifelse(x$chval > 999999, NA, x$chval)
> x$s20 <- ifelse(x$s20 > 999999, NA, x$s20)
> x$lwage_coap <- NA
> x[x$s31c > 0 & x$s31c < 999999,]$lwage_coap <-
+   log(x[x$s31c > 0 & x$s31c < 999999,]$s31c)
> x$lwage_coap2 <- ifelse(x$coap==1,x$lwage_coap,0)
> x$male_coap <- x$s16==1
```

We can add in a large number of variables that may be reasonably associated with legitimate mortgage denials, including measures of assets, ratio of debt to assets and property value. Lenders may also plausibly deny mortgages based on whether the mortgage has a co-applicant and characteristics of the co-applicant.

```
> W1 <- cbind(1,x$lwage,x$chist,x$mhist,x$phist,x$emp,
+              x$emp^2,x$married,x$dr,x$clines,x$male,
+              x$suff,x$assets,x$lr,x$pr,x$coap,x$s20,
+              x$s24a,x$s27a,x$s39,x$s48,x$s53,x$s55,x$s56,
+              x$s57,x$chval,x$school,x$bd,x$mi,x$old,
+              x$vr,x$uria,x$netw,x$dnotown,x$dprop,
+              x$lwage_coap2,x$lr^2,x$pr^2,x$clines^2,x$rtdum)
> # x$rtdum measures the racial make up of the neighborhood.
> index <- is.na(rowSums(cbind(Y1,X1,W1)))==0
> X2 <- X1[index,]
> Y2 <- Y1[index]
> W2 <- W1[index,]
```

2.6.5 Bootstrap Dual Path Estimator in R

The bootstrap estimator uses the algebra as pseudo-code for an estimator in **R**.

```
> set.seed(123456789)
> K <- 1000
> bs_mat <- matrix(NA,K,2)
> for (k in 1:K) {
+    index_k <- round(runif(length(Y2),min=1,max=length(Y2)))
+    Y3 <- Y2[index_k]
+    X3 <- X2[index_k,]
+    W3 <- W2[index_k,]
+    beta_tilde_hat <- solve(t(X3)%*%X3)%*%t(X3)%*%Y3
+    Delta_hat <- solve(t(X3)%*%X3)%*%t(X3)%*%W3
+    gamma_hat <- solve(t(W3)%*%W3)%*%t(W3)%*%Y3
+    bs_mat[k,] <- beta_tilde_hat - Delta_hat%*%gamma_hat
+    # print(k)
+ }
> tab_res <- matrix(NA,2,4)
> tab_res[,1] <- colMeans(bs_mat)
> tab_res[,2] <- apply(bs_mat, 2, sd)
> tab_res[1,3:4] <- quantile(bs_mat[,1], c(0.025,0.975))
> # first row, third and fourth column.
> tab_res[2,3:4] <- quantile(bs_mat[,2], c(0.025,0.975))
> colnames(tab_res) <- c("Estimate", "SD", "2.5%", "97.5%")
> rownames(tab_res) <- c("intercept","direct effect")
```

Adding in all these variables significantly reduces the estimate of the direct

	Estimate	SD	2.5%	97.5%
intercept	-0.00	0.00	-0.01	0.00
direct effect	0.02	0.02	-0.01	0.06

TABLE 2.7
Bootstrapped estimates of the proposed approach to estimating the direct effect of race on mortgage denials.

effect of race on mortgage denials. The estimated direct effect of being African American falls from a 20 percentage point reduction in the probability of getting a mortgage to a 2 percentage point reduction. A standard **hypothesis test** with bootstrapped standard errors states that we cannot rule out the possibility that the true direct effect of race on mortgage denials is zero.[15]

2.6.6   Policy Implications of Dual Path Estimates

The analysis shows that African Americans are much more likely to be denied mortgages in Boston during the time period. If this is something a policy maker wants to change, then she needs to know why African Americans are being denied mortgages. Is it direct discrimination of the banks? Or is the effect indirect because African Americans tend to have lower income and poorer credit ratings than other applicants. A policy that makes it illegal to use race directly in mortgage decisions will be more effective if bankers are in fact using race directly in mortgage decisions. Other policies may be more effective if bankers are using credit ratings and race is affecting loan rates indirectly.

Whether this analysis answers the question is left to the reader. It is not clear we should include variables such as the racial make up of the neighborhood or the gender of the applicant. The approach also relies on the assumption that there is in fact no direct effect of race on mortgage denials. In addition, it uses OLS rather than models that explicitly account for the discrete nature of the outcome variable.[16]

The approach presented here is quite different from the approach presented in Munnell et al. (1996). The authors are also concerned that race may have both direct and indirect causal effects on mortgage denials. Their solution is to estimate the relationship between the mediating variables (W) and mortgage denials (Y). In Figure 2.5 they are estimating c. The authors point out that the total effect of race on mortgage denials is not equal to c. They conclude that bankers must be using race directly. The problem is that they only measured part of the indirect effect. They did not estimate d. That is, their proposed methodology does not answer the question of whether bankers use race directly.

[15]See discussion of hypothesis testing in Appendix A.
[16]It may more appropriate to use a probit or logit. These models are discussed in Chapter 5.

2.7 Discussion and Further Reading

There is a simplistic idea that longer regressions must be better than shorter regressions. The belief is that it is *always* better to add more variables. Hopefully, this chapter showed that longer regressions can be better than shorter regressions, but they can also be worse. In particular, long regressions can create multicollinearity problems. While it is funny, Goldberger's chapter on multicollinearity downplays the importance of the issue.

The chapter shows that if we take DAGs seriously we may be able to use an alternative to the long regression. The chapter shows that in the case where there are two paths of a causal effect, we can improve upon the long regression. I highly recommend Pearl and Mackenzie (2018) to learn more about graphs. Pearl uses the term **mediation** to refer to the issue of dual causal paths.

To find out more about the lawsuit against Harvard University, go to https://studentsforfairadmissions.org/.

3

Instrumental Variables

3.1 Introduction

Despite being a relatively recent invention, the **instrumental variable** (IV) method has become a standard technique in microeconometrics.[1] The IV method provides a potential solution for one of the two substantive restrictions imposed by OLS. OLS requires that the unobserved characteristic of the individual enters into the model **independently** and **additively**. The IV method allows the estimation of **causal effects** when the independence assumption does not hold.

Consider again a policy in which Vermont makes public colleges free. Backers of such a policy believe that it will encourage more people to attend college. Backers also note that people who attend college earn more money. But do the newly encouraged college attendees earn more money?

Figure 3.1 presents median weekly earnings by education level for 2017. It shows that those with at least a Bachelor's Degree earn substantially more than those who never attended college. Over $1,100 per week compared to around $700. However, this is not the question of interest. The policy question is whether encouraging more people to attend college will lead *these* people to earn more. The problem is that people who attend college are different from people who do not attend college. Moreover, those differences lead to different earnings.

The chapter presents a general model of the IV estimator. It then presents a version of the estimator for **R**. The estimator is used to analyze returns to schooling and replicates Card (1995). The chapter uses the data to test whether the proposed instrumental variables satisfy the assumptions. The chapter ends with a discussion of an alternative IV estimator called the local average treatment effect (LATE) estimator. This estimator relaxes the additivity assumption.

---

[1] While becoming more prominent in the 1990s, instrumental variables is believed to date to the work of economist Philip Wright in the 1920s (Angrist and Krueger, 2001)

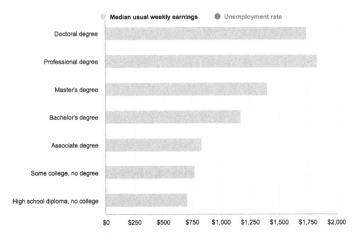

FIGURE 3.1
Chart of median weekly earnings by education level 2017 from Bureau of Labor Statistics.

3.2   A Confounded Model

The problem with comparing earnings from college attendees with those that have not attended college is **confounding**. There may be some unobserved characteristic associated with *both* attending college and earning income.

3.2.1   Confounded Model DAG

Consider the graph presented in Figure 3.2. We want to estimate the causal relationship between X and Y. X may represent college attendance and Y represents weekly earnings. The arrow from X to Y represents the causal effect of X on Y. This effect is denoted by b. Here, b, represents the increase in earnings due to a policy that increases the propensity to attend college. In order to evaluate the policy, we need to estimate b.

In Chapters 1 and 2 we saw a similar figure. This time we have an arrow from U to X. This extra arrow means the unobserved characteristics can determine the value of the policy variable. People who attend college may earn more than people who do not attend college for reasons that have nothing at all to do with attending college.

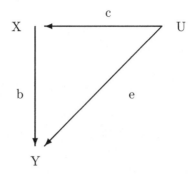

FIGURE 3.2
Confounded graph.

If our data is represented by Figure 3.2 we can try to estimate b by regressing Y on X. Doing so, however, will not give an estimate of b. Rather it will give an estimate of $b + \frac{e}{c}$. This is due to the **backdoor** problem. The issue is that there are two pathways connecting X to Y. One is the "front door" as represented by the arrow directly from X to Y with a weight of b. The second is the **backdoor**, which follows the backward arrow from X to U with the weight of $\frac{1}{c}$ and then the forward arrow from U to Y, with the weight of e.

3.2.2   Confounded Linear Model

We can write our confounded model with algebra.

$$
\begin{aligned}
y_i &= a + bx_i + ev_{1i} \\
x_i &= f + dz_i + v_{2i} + cv_{1i}
\end{aligned}
\tag{3.1}
$$

where y_i represents individual i's income, x_i is their education level, and v_{1i} and v_{2i} are unobserved characteristics that determine income and education level respectively. We are interested in estimating b. This is the same as the model we saw in the previous chapters. The difference here is that there is an additional equation.

To see the problem with OLS, rearrange the second equation and substitute it into the first equation.

$$
y_i = a + bx_i + e\left(\frac{x_i - f - dz_i - v_{2i}}{c}\right)
\tag{3.2}
$$

If we run OLS, we estimate the coefficient on x as $b + \frac{e}{c}$, not b. In Chapter 2 we looked to solve this problem by estimating e and c. We can't do that here because we do not observe v_{1i}.

3.2.3   Simulation of Confounded Data

```
> set.seed(123456789)
> N <- 1000
> a <- 2
> b <- 3
> c <- 2
> e <- 3
> f <- -1
> d <- 4
> z <- runif(N)
> u_1 <- rnorm(N, mean=0, sd=3)
> u_2 <- rnorm(N, mean=0, sd=1)
> # note we can change the mean and standard
> # deviation of the normal distribution
> x <- f + d*z + u_2 + c*u_1
> y <- a + b*x + e*u_1
> lm1 <- lm(y ~ x)
```

The simulated data is generated from the model presented above. The causal effect of x on y is $b = 3$. We see that v_1 affects both x and y. A standard method for estimating the causal effect of X on Y is to use OLS. Table 3.1 presents the OLS estimate of the simulated data. The table shows that the estimated relationship is 4.41. It is much larger in magnitude than the true value of 3. It is closer to the **backdoor** relationship of $b + \frac{e}{c} = 4.5$.

	Estimate	Std. Error	t value	Pr($>$|t|)
(Intercept)	0.5114	0.0706	7.24	0.0000
x	4.4090	0.0112	393.80	0.0000

TABLE 3.1
OLS estimate with confounded data.

We can estimate b, but our estimate is wrong.

3.3   IV Estimator

Perhaps the most common solution to confounding used in microeconometrics is the IV estimator. This section formally presents the estimator and illustrates it with the simulated data.

3.3.1 Graph Algebra of IV Estimator

Figure 3.3 represents the confounded data. It shows that the instrumental variable (Z) has a direct causal effect on X but is not determined by the unobserved characteristic. There is an arrow from Z to X and an arrow from U to X, but no arrow from U to Z.

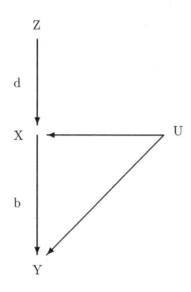

FIGURE 3.3
Experimental graph.

"Graph algebra" provides a solution. If we are interested in estimating b, we can do so by estimating the relationship between Z and Y. That effect is given by $b \times d$. We can estimate d by running a regression of X on Z. Thus dividing the result of the first regression by the result of the second regression gives the answer. If the unobserved characteristic affects the explanatory variable additively then the division procedure can be used to estimate the true causal relationship.

In the medical literature the first regression is called the **intent to treat** regression. The name comes from randomized controlled trials where patients are randomly assigned to alternative treatments, say a new cancer therapy versus the standard of care. Unfortunately, patients may not accept their treatment assignment. They may leave the trial or they may switch to the non-assigned treatment. In that case Z represents the randomization method while X represents the *actual* treatment received. By regressing Z on Y we are estimating causal effect of the initial randomized assignment. We are estimating the "intent" of the trial. Often only this estimate ($d \times b$) is presented, rather than effect of interest (b). See for example the analysis of commitment savings devices in the next chapter (Ashraf et al., 2006).

3.3.2 Properties of IV Estimator

In the standard analysis, an instrumental variable has three properties.

1. The variable directly affects the policy variable of interest ($Z \to X$).

2. The variable is independent of the unobserved characteristics that affect the policy variable and the outcome of interest ($U \not\to Z$).[2]

3. The variable affects the policy variable independently of the unobserved effect ($X = dZ + U$).

If we have a variable that satisfies these three properties, then we can use the graph algebra to estimate the causal effect of interest. Below we discuss how restrictive these assumptions are and test whether they hold in the data.

3.3.3 IV Estimator with Standard Algebra

A simple IV estimator is derived by rearranging the equations above (Equation (3.1)). We can determine the relationship between z_i and y_i by substituting x_i into the first equation.

$$y_i = a + b(f + dz_i + v_{2i} + cv_{1i}) + ev_{1i}$$
$$\text{or} \tag{3.3}$$
$$y_i = a + bf + bdz_i + bv_{2i} + bcv_{1i} + ev_{1i}$$

We can use OLS to give an unbiased estimate of the relationship between z_i and y_i, which is bd in the true model. The problem is that we are not interested in bd, we are interested in knowing b. Equation (3.1) shows we can use OLS to estimate the relationship between z_i and x_i. This gives an estimate of d. Thus, the division of the first estimate by the second gives an estimate of b.

3.3.4 Simulation of an IV Estimator

To illustrate the estimator, note that in the simulated data there exists a third variable z. This variable determines x but does not directly determine y and is independent of the unobserved characteristic (v).

Consider what happens if we run an OLS regression of y on z and a separate OLS regression of x on z.

```
> bd_hat <- lm(y ~ z)$coef[2]
> d_hat <- lm(x ~ z)$coef[2]
```

---

[2]The only relationship between the instrumental variable and the outcome of interest comes through the policy variable.

```
> # picking the slope coefficient from each regression
> bd_hat/d_hat
```

```
       z
3.015432
```

If we take the coefficient estimate from the first regression and divide that number by the coefficient estimate from the second regression, we get an estimate that is close to the true relationship. This is not coincidental. The observed characteristic, z, has the properties of an instrumental variable.

The procedure works well, but only if we want to instrument for one variable.

3.3.5 IV Estimator with Matrix Algebra

If we want to instrument for more than one variable we need matrix algebra. In matrix algebra, the model above is as follows.

$$\mathbf{y} = \mathbf{X}\beta + \upsilon \tag{3.4}$$

where $\mathbf{y}$ is a 100×1 vector of the outcome of interest y_i, $\mathbf{X}$ is a 100×2 matrix of the observed explanatory variables $\{1, x_i\}$, β is a 2×1 vector of the model parameters $\{a, b\}$ and υ is a 100×1 vector of the error term υ_i.

In addition, there is a relationship between the instrumental variables and the explanatory variables.

$$\mathbf{X} = \mathbf{Z}\Delta + \mathbf{E} \tag{3.5}$$

where $\mathbf{Z}$ is a 100×2 matrix of the instrumental variables $\{1, z_i\}$, Δ is a 2×2 matrix of the relationship between the explanatory variables and the instrumental variables, and $\mathbf{E}$ is a 100×2 matrix of unobserved characteristics determining the explanatory variables.

We can rearrange Equation (3.5) to determine Δ as a function of the observed and unobserved variables.

$$\Delta = (\mathbf{Z}'\mathbf{Z})^{-1}\mathbf{Z}'\mathbf{X} - (\mathbf{Z}'\mathbf{Z})^{-1}\mathbf{Z}'\mathbf{E} \tag{3.6}$$

Note that we use the exact same matrix procedure as we use for OLS in Chapter 1. There we "divided by" $\mathbf{X}$, here we "divide by" $\mathbf{Z}$.[3] The matrix $\mathbf{Z}'$ is the matrix transpose. By pre-multiplying we have a square matrix that may be invertible. It is assumed that the matrix $\mathbf{Z}$ is full-column rank.

A few more steps and we have our estimator. If we substitute Equation (3.5) into Equation (3.4) we have our intent to treat regression.

$$\mathbf{y} = \mathbf{Z}\Delta\beta + \mathbf{E}\beta + \upsilon \tag{3.7}$$

---

[3]If you are unfamiliar with matrix algebra I suggest looking at Kahn Academy or a matrix algebra text. The big difference between matrix algebra and regular algebra is that order matters. In addition, we can only invert full-rank square matrices.

Rearranging this equation we can get an estimator for the coefficients ($\Delta\beta$).

$$\Delta\beta = (\mathbf{Z'Z})^{-1}\mathbf{Z'y} - (\mathbf{Z'Z})^{-1}\mathbf{Z'E}\beta - (\mathbf{Z'Z})^{-1}\mathbf{Z'}\upsilon \qquad (3.8)$$

Substituting Equation (3.6) into this equation and simplifying, we have the following relationship.

$$\beta = (\mathbf{Z'X})^{-1}\mathbf{Z'y} - (\mathbf{Z'X})^{-1}\mathbf{Z'}\upsilon \qquad (3.9)$$

From Equation (3.9) we have our instrumental variable estimator.

$$\hat{\beta}_{IV} = (\mathbf{Z'X})^{-1}\mathbf{Z'y} \qquad (3.10)$$

Remember, we don't observe the last part of Equation (3.9). This means that our estimate of $\hat{\beta}_{IV}$ will not be equal to β. The question is when they are likely to be close to each other.[4]

This generalizes OLS. If the instrumental variable matrix ($\mathbf{Z}$) is equal to the explanatory variable matrix ($\mathbf{X}$) then this is just our OLS estimator. This relationship makes it clear that OLS assumes that each explanatory variable is independent of the unobserved characteristic. There is no arrow from U to X. More generally, the matrix of instrumental variables must be the same size as the matrix of explanatory variables. Each explanatory variable must either have an assigned instrument or must be an instrument for itself.[5]

3.3.6   Two-Stage Least Squares

A common algorithm for IV is two-stage least squares. This algorithm operationalizes the matrix algebra steps above. From Equation (3.6) we have a first-stage estimator.

$$\hat{\Delta} = (\mathbf{Z'Z})^{-1}\mathbf{Z'X} \qquad (3.11)$$

That is, we regress the set of instruments (exogenous variables) on the set of endogenous variables. Note that the result is a matrix, rather than a vector in the standard OLS model.

We can use this estimate in the second-stage regression. We can go to Equation (3.7) and replace the $\mathbf{Z}\Delta$ with $\mathbf{Z}\hat{\Delta}$. That is, we replace the endogenous variables with their predicted values.

I have not coded up the two-stage least squares algorithm. We use the algorithm derived from the matrix algebra. Can you code the two-stage least squares estimator? How does it compare to the estimator below?

---

[4]This question is discussed in more detail in Appendix A.

[5]Here we have a maximum of one instrument for each endogenous variable. In Chapter 8 we allow multiple instruments for each endogenous variable using **generalized method of moments**.

3.3.7   IV Estimator in R

We can use Equation (3.10) as **pseudo-code** for the instrumental variable estimator. Note that while the OLS estimator gives estimates that diverge substantially from the true values for both the intercept and the slope, the IV estimates are a lot closer to the true values.

```
> X <- cbind(1,x) # remember the column of 1's for the intercept
> Z <- cbind(1,z) # remember Z same size as X
> beta_hat_ols <- solve(t(X)%*%X)%*%t(X)%*%y
> beta_hat_iv <- solve(t(Z)%*%X)%*%t(Z)%*%y
> beta_hat_ols

        [,1]
  0.5113529
x 4.4090257

> beta_hat_iv

        [,1]
  1.715181
x 3.015432
```

3.3.8   Bootstrap IV Estimator for R

The following **bootstrap** IV estimator defaults to an OLS estimator. It uses the matrix algebra above to create a function that we can use on various problems. In **R** the function() operator is used to create an object that takes inputs and produces an output. In this case, it takes in our data and produces an instrumental variable estimate.

To use this estimator, you can copy the code and then "run" that proportion of the code defining the function. Once that is done, you can call the function in your **R** script.

```
> lm_iv <- function(y, X_in, Z_in = X_in, Reps = 100,
+                    min_in = 0.05, max_in = 0.95) {
+    # takes in the y variable, x explanatory variables
+    # and the z variables if available.
+    # The "=" creates a default value for the local variables.
+
+    # Set up
+    set.seed(123456789)
+    X <- cbind(1,X_in)  # adds a column of 1's the matrix
+    Z <- cbind(1,Z_in)
+
+
+    # Bootstrap
```

```
+    bs_mat <- matrix(NA,Reps,dim(X)[2])
+    # dim gives the number of rows and columns of a matrix.
+    # the second element is the number of columns.
+    N <- length(y) # number of observations
+    for (r in 1:Reps) {
+       index_bs <- round(runif(N, min = 1, max = N))
+       y_bs <- y[index_bs]  # note Y is a vector
+       X_bs <- X[index_bs,]
+       Z_bs <- Z[index_bs,]
+       bs_mat[r,] <- solve(t(Z_bs)%*%X_bs)%*%t(Z_bs)%*%y_bs
+    }
+
+    # Present results
+    tab_res <- matrix(NA,dim(X)[2],4)
+    tab_res[,1] <- colMeans(bs_mat)
+    for (j in 1:dim(X)[2]) {
+       tab_res[j,2] <- sd(bs_mat[,j])
+       tab_res[j,3] <- quantile(bs_mat[,j],min_in)
+       tab_res[j,4] <- quantile(bs_mat[,j],max_in)
+    }
+    colnames(tab_res) <- c("coef","sd", as.character(min_in),
+                                 as.character(max_in))
+    return(tab_res)
+    # returns the tab_res matrix.
+ }
```

The function takes as inputs the vector of outcome variables (y), a matrix of explanatory variables (the Xs), and a matrix of instruments (Zs).[6] It defaults to allowing $\mathbf{X} = \mathbf{Z}$. It also assumes that the matrix of instruments is the exact same dimension as the matrix of explanatory variables. The function then tidies up by adding the column of 1s to both the matrix of explanatory variables and the matrix of instruments. The function accounts for sampling error by using the bootstrap methodology.[7] The bootstrap creates pseudo-random samples by re-sampling the original data and re-estimating the model on each new pseudo-sample. In each case, it uses the matrix algebra presented above to calculate the estimator.

```
> print(lm_iv(y,x), digits = 3)  # OLS

        coef       sd   0.05   0.95
[1,] 0.513 0.0729 0.396 0.643
[2,] 4.409 0.0109 4.394 4.427

> print(lm_iv(y,x,z), digits = 3) # IV
```

---

[6] An alternative, more robust, version of this function is presented in Appendix B.
[7] See discussion in Chapter 1 and Appendix A.

```
     coef     sd 0.05 0.95
[1,] 1.83 0.377 1.30 2.43
[2,] 2.96 0.286 2.45 3.37
```

We can compare the IV and OLS estimators on the simulated data. Using the IV estimator with the simulated data shows that the OLS estimates are biased and that the true values do not lie within the 90% confidence interval. The IV estimates do better.

3.4 Returns to Schooling

Chapters 1 and 2 discussed how Canadian labor economist, David Card, used OLS to estimate returns to schooling. Card finds that an extra year of schooling increases income by approximately 7.5%. Card points out that this estimate may be biased and argues for using the instrumental variable approach.

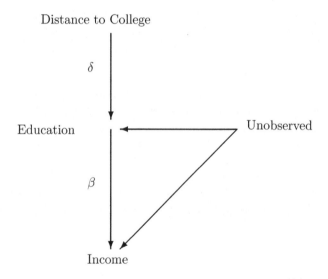

FIGURE 3.4
Returns to schooling graph.

The unobserved characteristics of the young men may determine both the amount of education that they get and the income they earn. In Figure 3.4, the problem is illustrated with a causal arrow running from the unobserved characteristic to both income and education. For example, young men from wealthier families may be more likely to go to college because their families can afford it. In addition, these young men may go into well paying jobs due to family connections. In this case, the observed positive relationship we see between education and income may be due to the wealth of the young man's family rather than the education itself. The effect of the confounding is that we get an inaccurate estimate of β. If a policy is designed to increase education it won't have the expected effect on income.

3.4.1   Distance to College as an Instrument

Card (1995) argues that young men who grow up near a 4 year college will have lower costs to attending college and are thus more likely to get another year of education. In addition, Card argues that growing up close to a 4 year college is unlikely to be determined by unobserved characteristics that also determine the amount of education that the young man gets and the income that the young man earns. In the graph, the assumption is represented as an arrow from "distance to college" to education and *no* arrow from unobserved characteristics to "distance to college."

Formally, the model used in Card (1995) is as follows.

$$\log \text{wage76}_i = \alpha_1 + \beta \; \delta \; \text{nearCollege}_i + \gamma_1 \; \text{observables}_i + \text{unobservables}_{i1}$$

$$ed_i = \alpha_2 + \delta \; \text{nearCollege}_i + \gamma_2 \; \text{observables}_i + \text{unobservables}_{i2}$$

$$(3.12)$$

In Equation (3.12) return to schooling is given by the β parameter, which measures the impact of an additional year of schooling on log wages. If we are willing to make the independence and additivity assumptions discussed above, then we can use IV to estimate returns to schooling.

We estimate two regressions. The first, the intent to treat estimate, is the estimated effect on the outcome of interest by the instrumental variable. The estimated effect is made up of two effects, the return to schooling effect (β) and the effect of the instrumental variable on the propensity to get another year of education (δ). The second regression estimates this second part. Our estimate of returns to schooling comes from taking the first estimate and dividing it by the second.

3.4.2   NLSM Data

Card (1995) uses the National Longitudinal Survey of Older and Younger Men (NLSM) data. The data set used by David Card can be downloaded from his website. Chapter 1 discusses how to download the data and Chapter 2 discusses how to create the variables needed.[8]

---

[8]The version of the data used here can be found here: https://sites.google.com/view/microeconometricswithr/table-of-contents.

```
> x <- read.csv("nls.csv",as.is=TRUE)
> x$lwage76 <- as.numeric(x$lwage76)
> x1 <- x[is.na(x$lwage76)==0,]
> x1$exp <- x1$age76 - x1$ed76 - 6 # working years after school
> x1$exp2 <- (x1$exp^2)/100 # experienced squared divided by 100
```

3.4.3   Simple IV Estimates of Returns to Schooling

Using the NLSM data set, we can follow the procedure laid out above. First, for comparison purposes, estimate the OLS model of returns to schooling. Then estimate the intent to treat equation of the instrument, distance to college, on income. Then estimate the effect of distance on education. Finally, use division to determine the causal effect of education on income.

```
> # OLS Estimate
> lm4 <- lm(lwage76 ~ ed76 + exp + exp2 + black + reg76r +
+                smsa76r + smsa66r + reg662 + reg663 + reg664 +
+                reg665 + reg666 + reg667 + reg668 + reg669,
+           data=x1)
> # smsa refers to urban or rural, with 76 refering to 1976.
> # reg refers to region of the US - North, South, West etc.
> # 66 refers to 1966.
> lm4$coefficients[2]

      ed76
0.07469326

> # Intent To Treat Estimate
> lm5 <- lm(lwage76 ~ nearc4 + exp + exp2 + black + reg76r +
+                smsa76r + smsa66r + reg662 + reg663 + reg664 +
+                reg665 + reg666 + reg667 + reg668 + reg669,
+           data=x1)
> # nearc4 is a dummy for distance to a 4 year college.
> lm5$coefficients[2]

    nearc4
0.04206793

> # Effect of instrument on explanatory variable
> lm6 <- lm(ed76 ~ nearc4 + exp + exp2 + black + reg76r +
+                smsa76r + smsa66r + reg662 + reg663 + reg664 +
ı                reg665 + reg666 + reg667 + reg668 + reg669,
+           data=x1)
> lm6$coefficients[2]

   nearc4
0.3198989
```

```
> # IV Estimate of Returns to Schooling
> lm5$coefficients[2]/lm6$coefficients[2]

  nearc4
0.1315038
```

The estimate is much larger than suggested by the OLS model. The β from the OLS estimate is around 0.074, while the β from the IV estimate is around 0.132. Why is that?[9]

3.4.4   Matrix Algebra IV Estimates of Returns to Schooling

Card (1995) uses multiple instruments. The paper also instruments for experience. Experience is measured as the difference between age and years of education, but education is confounded. The paper uses age as the instrument for experience and age-squared as the instrument for experience-squared.

The instrumental variable procedure derived above is used to estimate the returns to schooling accounting for all three instruments.

```
> y <- x1$lwage76
> X <- cbind(x1$ed76, x1$exp, x1$exp2, x1$black, x1$reg76r,
+            x1$smsa76r, x1$smsa66r, x1$reg662, x1$reg663,
+            x1$reg664, x1$reg665, x1$reg666, x1$reg667,
+            x1$reg668, x1$reg669)
> x1$age2 <- x1$age76^2
> Z1 <- cbind(x1$nearc4, x1$age76, x1$age2, x1$black,
+             x1$reg76r,x1$smsa76r, x1$smsa66r, x1$reg662,
+             x1$reg663,x1$reg664, x1$reg665, x1$reg666,
+             x1$reg667, x1$reg668, x1$reg669)
> res <- lm_iv(y,X,Z1, Reps=1000)
> rownames(res) <- c("intercept","ed76","exp", "exp2", "black",
+                 "reg76r","smsa76r", "smsa66r",
+                 "reg662","reg663","reg664", "reg665",
+                 "reg666","reg667", "reg668", "reg669")
```

The results replicate those in Card (1995). Table 3.2 can be compared to the original estimates in Table 3 of the paper (Card, 1995). The main difference is that the estimated standard errors from the bootstrap are much higher than from the method used to calculate them in Table 3.

The instrumental variable estimate on returns to schooling is substantially higher than the OLS estimate. This is the opposite of what I expected. I expected the OLS estimates of the causal effect of schooling on income to be biased up. The concern is that OLS is picking up the fact that family background is determining both college attendance and access to high paying

---

[9]This issue is discussed further below.

	coef	sd	0.05	0.95
intercept	3.98	0.82	2.78	4.95
ed76	0.13	0.07	0.05	0.24
exp	0.06	0.04	0.01	0.10
exp2	-0.10	0.19	-0.32	0.16
black	-0.12	0.10	-0.24	0.03
reg76r	-0.14	0.03	-0.19	-0.09
smsa76r	0.08	0.08	-0.04	0.17
smsa66r	0.03	0.02	-0.00	0.07
reg662	0.08	0.05	0.00	0.15
reg663	0.13	0.05	0.05	0.20
reg664	0.03	0.07	-0.06	0.12
reg665	0.14	0.05	0.06	0.22
reg666	0.15	0.05	0.07	0.23
reg667	0.12	0.05	0.04	0.21
reg668	-0.10	0.09	-0.23	0.01
reg669	0.09	0.06	-0.00	0.17

TABLE 3.2
A replication of some results from Part (B) of Table 3 from Card (1995).

jobs. The results suggest the OLS estimate is biased down. It is unclear why this would be. The rest of the chapter discusses the assumptions of the IV model and how we may test them and weaken them.

3.4.5 Concerns with Distance to College

There are two major concerns with this approach. The first concern is that distance to college is not an instrumental variable. This occurs if unobserved characteristics associated with living near a college are associated with higher earnings. To test this we can compare observable characteristics for those who grew up near a college and those that did not.

```
> tab_cols <- c("Near College", "Not Near College")
> tab_rows <- c("ed76","exp","black","south66",
+               "smsa66r","reg76r","smsa76r")
> # these are the variables we are interested in comparing.
> table_dist <- matrix(NA,7,2)
> # loop creating mean of each variable for each type
> for (i in 1:7) {
+    table_dist[i,1] <-
+      mean(x1[x1$nearc4==1,colnames(x1)==tab_rows[i]])
+    table_dist[i,2] <-
+      mean(x1[x1$nearc4==0,colnames(x1)==tab_rows[i]])
+ }
```

```
> colnames(table_dist) <- tab_cols
> rownames(table_dist) <- tab_rows
```

Table 3.3 presents a comparison of means between the two groups. As expected and required, the group that grew up near a college has more education and less work experience by 1976. However, they also seem to be systemically different in other observable ways. This group is less likely to be black, less likely to live in the south and more likely to live in a city. All these are positively associated with earning more income. The table suggests that distance to college does not satisfy the assumptions of an instrumental variable.

	Near College	Not Near College
ed76	13.53	12.70
exp	8.68	9.23
black	0.21	0.28
south66	0.33	0.60
smsa66r	0.80	0.33
reg76r	0.33	0.56
smsa76r	0.82	0.48

TABLE 3.3
Mean values for students that did not grow up near a college and students that did grow up near a college.

The second concern is the additivity assumption. We assumed that the instrumental variable affects the explanatory variable additively. The assumption allows the unobserved characteristics of the student and their distance to college to determine the amount of schooling they receive. It does not allow the two effects to interact. For example, for students from families with less means, living near college may have a big effect on their propensity to go to college. However, for students from wealthy families, it has little or no effect. This concern is discussed in more detail in the last section of the chapter.

3.5   Instrument Validity

The instrumental variable method is a powerful and popular algorithm. To use it, we need an observed characteristic that satisfies three assumptions. First, the observed characteristic needs to causally affect the policy variable. Second, the observed characteristic is not affected by unobserved characteristics affecting the outcome of interest. Third, the observed characteristic's effect on the policy variable is additively separable from the unobserved characteristic's effect. How do we know if an instrument satisfies the assumptions of the model? Often we don't.

In some cases we can test whether instrumental variable assumptions hold.

3.5.1 Test of Instrument Validity

If we have two potential instruments we can test the model. Consider the model represented by the graph in Figure 3.5. We have two instruments, Z_1 and Z_2. Note that these two instruments give us two different estimates of b. Let $c_1 = d_1 \times b$ and $c_2 = d_2 \times b$. By regressing Z_1 on Y we can estimate c_1. Similarly, by regressing Z_2 on Y we can estimate c_2. Separately, we can also estimate d_1 and d_2.

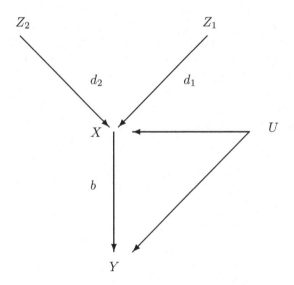

FIGURE 3.5
Two instrument graph.

These two sets of regressions give us an **over-identification test**.[10] We can ask if our two ways of estimating b give the exactly the same answer.

$$\left| \frac{c_1}{d_1} - \frac{c_2}{d_2} \right| = 0 \qquad (3.13)$$

The test is illustrated using Card's returns to schooling data. Assume that we have two valid instruments, distance to a 4 year college and whether both

---

[10] A model is over identified if we can estimate the parameter more than one way.

parents were at home when the young man was 14 (Kédagni, 2017). It is
assumed that an individual growing up near a 4 year college is more likely to
go to college and an individual growing up with both parents is more likely to
go to college. In addition, it is assumed that the unobserved characteristics of
the individual that affect their propensity to go to college and their earnings
in 1976 do not affect these two measures.

3.5.2 Test of Instrument Validity in R

Using the matrix algebra approach, we can compare the two alternative
estimates for returns to schooling. If the model represented by Figure 3.5 holds,
then the difference between the two estimates is zero. To test this we can run
the two different estimates on pseudo-random samples following a bootstrap
procedure.[11]

```
> Z2 <- cbind(x1$momdad14, x1$age76, x1$age2, x1$black,
+              x1$reg76r, x1$smsa76r, x1$smsa66r, x1$reg662,
+              x1$reg663, x1$reg664, x1$reg665, x1$reg666,
+              x1$reg667, x1$reg668, x1$reg669)
> # Bootstrap
> set.seed(123456789)
> bs_diff <- matrix(NA,1000,1)
> N <- length(y)
> for (i in 1:1000) {
+    index_bs <- round(runif(N, min = 1, max = N))
+    y_bs <- y[index_bs]
+    X_bs <- X[index_bs,]
+    Z1_bs <- Z1[index_bs,]
+    Z2_bs <- Z2[index_bs,]
+    bs_diff[i,] <-
+      (solve(t(Z1_bs)%*%X_bs)%*%t(Z1_bs)%*%y_bs)[2,1] -
+      (solve(t(Z2_bs)%*%X_bs)%*%t(Z2_bs)%*%y_bs)[2,1]
+    # note the parentheses around the beta estimates.
+    # print(i)
+ }
> summary(bs_diff)

        V1
 Min.   :-30.00616
 1st Qu.: -0.03099
 Median :  0.20354
 Mean   :  0.67618
 3rd Qu.:  0.51374
 Max.   :275.62954
```

---

[11]Note that both should be estimated on the same pseudo-random sample.

```
> quantile(bs_diff,c(0.05,0.95))
```

```
       5%       95%
-0.3744408  1.7518602
```

The mean difference is about 0.68 and the 90% confidence interval includes zero. Under the standard hypothesis testing approach we cannot rule out that the model assumptions hold. We cannot rule out that distance to college and having both parents at home are both valid instruments.

Warning! Note the carefully parsed sentence above. It does not say that distance to college is a valid instrument. The test results do not state that we have a valid instrument. In fact, there is no circumstance in which they would. There are circumstances where the results allow us to say that one or both of the instruments are *invalid*. But, that is not the case here.

The results tell us clearly that the instrument may be valid or may be invalid.

3.6 Better LATE than Nothing

The assumption that the instrumental variable affects the policy variable independently of the unobserved characteristic may be more concerning than the other assumptions. In some cases, we can drop this restriction and simply reinterpret the result.

This section discusses the problem and presents a solution, the local average treatment effect (LATE).

3.6.1 Heterogeneous Effects

It is not reasonable to assume that the policy has the *same* effect on everyone. Introspection suggests that some people get more out of attending college than others. If we introduced a policy that encouraged more people to attend college, it is unlikely that each person would get the same benefit. It is even possible that some people would actually earn less income going to college than if they had not gone. That is, for some people the treatment effect may be negative. College may **cause** their income to fall.

The treatment effect may be **heterogeneous**. Nevertheless we may be interested in the **average treatment effect**. For example, if we have a policy that forces everyone to get an additional year of schooling, the average treatment effect may give a reasonable measure of the policy's effect on income.[12] Note that the terminology comes from statistics where we think of estimating the effect of giving a cancer patient a new drug treatment. Usually, we compare

---

[12]Most places do in fact have laws requiring school attendance to certain ages.

the new drug treatment to the "standard of care." The measured difference in survival rates between the new drug and the standard of care is the treatment effect. The addition of the word "average" suggests that the treatment effect varies across patients.

If the treatment effect is heterogeneous, the instrumental variable approach is not valid. The approach does not allow the instrumental variable to systematically affect the policy variable in a way that is related to the outcome. We cannot measure the average treatment effect if the U and Z interact in affecting X. But all may not be lost. In some cases we can interpret the estimate as the average effect for a subset of the population (Card, 2001). This estimand is called the **Local Average Treatment Effect** or LATE. The Stanford econometrician, Guido Imbens, argues that LATE is better than nothing (Imbens, 2010).

3.6.2   Local Average Treatment Effect

Assume there are four groups of people. These four groups are characterized by the probability that they accept the treatment corresponding to the instrument. Importantly, we do not observe which group a particular person is in.

1. Compliers: $\Pr(X = 1|Z = 1, C) = \Pr(X = 0|Z = 0, C) = 1$
2. Always Takers: $\Pr(X = 1|Z = 1, A) = \Pr(X = 1|Z = 0, A) = 1$
3. Never Takers: $\Pr(X = 0|Z = 1, N) = \Pr(X = 0|Z = 0, N) = 1$
4. Defiers: $\Pr(X = 0|Z = 1, D) = \Pr(X = 1|Z = 0, D) = 1$

Both the policy variable (X) and the instrument (Z) are either 0 or 1. In the returns to schooling example, X may represent college attendance while Z represents whether the young man lives close to a college or not.

The four types are compliers, always takers, never takers and defiers. A complier is a person who attends college $(X = 1)$ if they live close to a college $(Z = 1)$ but does not attend college $(X = 0)$ if they live far from a college $(Z = 0)$. This group "complies" with the treatment assignment. An always taker attends college $(X = 1)$ irrespective of where they live $(Z \in \{0, 1\})$. A never taker, never attends college $(X = 0)$. Finally, a defier attends college $(X = 1)$ if they live far from a college $(Z = 0)$, but does not attend college $(X = 0)$ if they live close to a college $(Z = 1)$.

There is no expectation that these groups are immutable. Rather they are determined by the economics. Distance to college is a "price." It is part of the price or cost of attending a particular college. If the price changes, demand changes. However, only some people change their demand for the product when the price changes. Some people keep buying the product at the new price and some people never buy the product, no matter what the price is. That said, few people buy more when the price increases.

For each type, we can write down the intent to treat effect following the

Law of Total Expectation.[13]

$$\mathbb{E}(Y|Z=1) - \mathbb{E}(Y|Z=0)$$
$$=$$
$$\sum_{T\in\{C,A,N,D\}} \left(\mathbb{E}(Y|Z=1,T) - \mathbb{E}(Y|Z=0,T)\right)\Pr(T) \tag{3.14}$$

where T represents one of our four groups. The unconditional intent to treat is a weighted average of the intent to treat for each group, weighted by each group's fraction of the population.

We can write out expected outcome conditional on the instrument and the type via the same law.

$$\mathbb{E}(Y|Z=1,T) \ = \mathbb{E}(Y|Z=1,X=1,T)\Pr(X=1|Z=1,T)$$
$$+\mathbb{E}(Y|Z=1,X=0,T)\Pr(X=0|Z=1,T) \tag{3.15}$$

The expected income conditional on the instrument is an average of the expected income conditional on both the instrument and the treatment allocation, weighted by the probability of receiving the treatment allocation conditional on the instrument.

The effect of Z on Y is *only* through X. By definition, once we know the value of the policy variable, the value of the instrument is irrelevant. Under the assumption that Z is an instrument we have that $\mathbb{E}(Y|X=1,Z=1) = \mathbb{E}(Y|X=1,Z=0) = \mathbb{E}(Y|X=1)$.

This implies the following for our intent to treat estimates for each group.

$$\mathbb{E}(Y|Z=1,C) - \mathbb{E}(Y|Z=0,C) = \mathbb{E}(Y|X=1,C) - \mathbb{E}(Y|X=0,C)$$
$$\mathbb{E}(Y|Z=1,A) - \mathbb{E}(Y|Z=0,A) = \mathbb{E}(Y|X=1,A) - \mathbb{E}(Y|X=1,A) = 0$$
$$\mathbb{E}(Y|Z=1,N) - \mathbb{E}(Y|Z=0,N) = \mathbb{E}(Y|X=0,N) - \mathbb{E}(Y|X=0,N) = 0$$
$$\mathbb{E}(Y|Z=1,D) - \mathbb{E}(Y|Z=0,D) = \mathbb{E}(Y|X=0,D) - \mathbb{E}(Y|X=1,D)$$
$$\tag{3.16}$$

For two of the types the intent to treat regression is zero. That leaves the compliers and defiers. Given the additional assumption that there are no defiers $(\Pr(D)=0)$ (also called a monotonicity assumption), we observe the intent to treat for compliers.

Given that we can observe the fraction of compliers, we have the result

$$\mathbb{E}(Y|X=1,C) - \mathbb{E}(Y|X=0,C) = \frac{\mathbb{E}(Y|Z=1) - \mathbb{E}(Y|Z=0)}{\Pr(X=1|Z=1) - \Pr(X=1|Z=0)}$$
$$\tag{3.17}$$

Note the value on the bottom of the fraction is the percent of compliers. Only always takers and compliers attend college, and the number of always takers does not change with the value of the instrument. Any change in the college attendance associated with a change in the value of the instrument is then due to compliers.

---

[13] We can always write out probability of an event as a weighted sum of all the conditional probabilities of the event. That is, $\Pr(A) = \Pr(A|B)\Pr(B) + \Pr(A|C)\Pr(C)$, where $\Pr(B) + \Pr(C) = 1$.

This fraction is the discrete version of the IV estimate presented above. The LATE estimate is an alternative interpretation of the original estimate.

3.6.3 LATE Estimator

It is straightforward to derive the LATE estimator from Equation (3.17). It is simply the **intent to treat** divided by the effect of Z on X.

The empirical analog of the top of the fraction is as follows.

$$\hat{\mu}_{y1} = \frac{\sum_{i=1}^{N} y_i \mathbb{1}(z_i = 1)}{\sum_{i=1}^{N} \mathbb{1}(z_i = 1)} \tag{3.18}$$

and

$$\hat{\mu}_{y0} = \frac{\sum_{i=1}^{N} y_i \mathbb{1}(z_i = 0)}{\sum_{i=1}^{N} \mathbb{1}(z_i = 0)} \tag{3.19}$$

where $\mathbb{1}()$ is an indicator function. This function is 1 if the value inside the parenthesis is true, 0 if it is false.

We can also write out the analog estimators for the two bottom probabilities.

$$\hat{p}_{11} = \frac{\sum_{i=1}^{N} \mathbb{1}(x_i = 1 \ \& \ z_i = 1)}{\sum_{i=1}^{N} \mathbb{1}(z_i = 1)} \tag{3.20}$$

and

$$\hat{p}_{10} = \frac{\sum_{i=1}^{N} \mathbb{1}(x_i = 1 \ \& \ z_i = 0)}{\sum_{i=1}^{N} \mathbb{1}(z_i = 0)} \tag{3.21}$$

Putting all this together, we have the LATE estimator.

$$\hat{\mu}_{LATE} = \frac{\hat{\mu}_{y1} - \hat{\mu}_{y0}}{\hat{p}_{11} - \hat{p}_{10}} \tag{3.22}$$

3.6.4 LATE Estimates of Returns to Schooling

To use the LATE approach we need to make education a binary variable. This is done by splitting the population between those that go to college and those that don't. It is assumed that everyone with more than 12 years of education goes to college.

The LATE estimator derived above is coded below. Note that mean() of a binary variable gives the probability.

```
> X2 <- X[,1] > 12  # college indicator
> # using college proximity as an instrument.
> mu_y1 <- mean(y[Z1[,1]==1])
> mu_y0 <- mean(y[Z1[,1]==0])
> p_11 <- mean(X2[Z1[,1]==1])
> p_10 <- mean(X2[Z1[,1]==0])
```

```
> # LATE, divide by 4 to get the per-year effect
> ((mu_y1 - mu_y0)/(p_11 - p_10))/4

[1] 0.3196679

> # this allows comparison with the OLS estimates.
>
> # using living with both parents as an instrument.
> mu_y1 <- mean(y[Z2[,1]==1])
> mu_y0 <- mean(y[Z2[,1]==0])
> p_11 <- mean(X2[Z2[,1]==1])
> p_10 <- mean(X2[Z2[,1]==0])
> ((mu_y1 - mu_y0)/(p_11 - p_10))/4

[1] 0.1772967
```

We can compare the LATE for the two proposed instruments. The estimate of the average annual effect of attending college is 0.32 using college proximity. It is 0.18 using living with both parents as an instrument. These estimates are both much larger than the OLS estimates. Note that we have not controlled for other observed characteristics of the individual including age, race etc. Table 3.3 shows that those individuals growing up near colleges have a number of other observed characteristics associated with higher incomes.

The variation in the LATE estimates suggests that returns to schooling is heterogeneous. Those whose college attendance is affected by their distance to college get very high returns. While those who are affected by having both parents at home get lower returns to college. See Kédagni (2017) for a discussion of this issue.

3.7 Discussion and Further Reading

IV has become a standard technique in microeconometrics. It allows the researcher to weaken the assumption that the policy variable is **independent** of unobserved characteristics.

Using OLS can lead to biased estimates. There maybe a **backdoor relationship** between the policy variable and the outcome variable of interest. In the previous chapter, we solved the bias by directly estimating the backdoor path. We cannot use that method here because we don't observe the unobserved characteristic. Instead, we can use instrumental variables.

Instrumental variables can be thought of as a "randomization device." They cause individuals to get more or less education for reasons not related to the income they will receive. Using instrumental variables, we can estimate the causal effect of the policy variable on the outcome of interest in two steps.

In the first step we estimate the **intent to treat regression**. That is, we regress the outcome of interest on the **instrumental variable**. This gives us the wrong answer, but we can get the right answer by dividing the intent to treat by the result of a regression of the policy variable on the instrument. The chapter shows that this "division" idea can be generalized to multiple variables using matrix algebra.

Card (1995) argues that we can get unbiased estimates of returns to schooling by using "distance to college" as an instrument. The argument is that living closer to a college reduces the cost of attending college and thus makes it more likely the person will attend. The IV estimates of returns to schooling are about twice the OLS estimates.

IV requires strong assumptions and researchers need to take care when choosing instruments. The chapter presents one way to test instrument validity. It also presents the idea of Local Average Treatment Estimation (LATE). LATE allows the econometrican to relax the assumption that the unobserved characteristic has an additive effect on the outcome of interest. However, it is unclear whether the LATE estimand provides the policy maker with valuable information. Imbens (2010) presents one side of the debate on the value of LATE. Kédagni (2017) considers what can be said when the IV assumptions are relaxed.

The next chapter discusses an alternative approach without the strong assumptions of the IV method. Chapter 6 presents a related approach using to estimating returns to schooling, the Heckman selection model. Chapter 8 presents the generalized method of moments approach to IV estimation of returns to schooling.

4

Bounds Estimation

4.1 Introduction

In the first three chapters we estimated, or attempted to estimate, a single value of interest. This chapter considers situations where we are either unable or unwilling to estimate a single value for the policy parameter of interest. Instead, the chapter considers cases where we are limited to estimating a range of values. We are interested in using the data to estimate the **bounds** on the policy parameter of interest.

It is standard practice in econometrics to present the **average treatment effect** (ATE). This estimand provides the policy maker with the average impact of the policy if *everyone* was to receive the policy. That is, if everyone changes from not attending college to attending college, the ATE predicts what would happen. I would give an example, but I can't think of one. In general, policies do not work like this. Consider the policy of making public state colleges free. Such a policy would encourage more people to attend college, but a bunch of people were already attending college and a bunch of people will not attend college even if it is free. What does the ATE tell us will happen to those that are newly encouraged to go to college? Not that much.

If attending college has the same effect on everyone, then the ATE provides useful information. If everyone has the same treatment effect, the average must be equal to the treatment effect. The difficulty arises when different people get different value from going to college. That is, the difficulty always arises.

This chapter considers two implications. In the first case, the data allows the ATE to be estimated, but we would prefer to know the distribution of the policy effect. In general, we cannot estimate this distribution. We can, however, bound it. These bounds are based on a conjecture of the great Soviet mathematician, Andrey Kolmogorov. The chapter explains how the **Kolmogorov bounds** work and when they provide the policy maker with useful information. These bounds are illustrated by analyzing a randomized controlled trial on the effect of "commitment savings" devices.

In the second case, the data does not allow the ATE to be estimated. Or more accurately, we are unwilling to make the non-credible assumptions necessary to estimate the ATE. The Northwestern econometrician, Charles Manski, argues that econometricians are too willing to present estimates based on non-credible assumptions. Manski shows that weaker but more credible assumptions often lead to a range of estimates. He suggests that presenting a range of estimates is better than providing precisely estimated nonsense. The chapter presents Manski's **natural bounds** and discusses how assumptions can reduce the range of estimates of the policy effect. The chapter illustrates these ideas by estimating whether more guns reduce crime.

4.2   Potential Outcomes

You have been tasked by the AOC 2028 campaign to estimate the likely impact of a proposal to make state public universities tuition free.[1] Your colleague is tasked with estimating how many more people will attend college once it is made free. You are to work out what happens to incomes of those that choose to go to college, now that is free. You need to estimate the **treatment effect** of college.

4.2.1   Model of Potential Outcomes

Consider a simple version of the problem. There are two possible outcomes. There is the income the individual receives if they attend college $(y_i(1))$ and the income they would receive if they did not attend college $(y_i(0))$.

$$y_i(x_i) = a + b_i x_i + v_i \qquad (4.1)$$

where y_i is individual i's income, $x_i \in \{0, 1\}$ is whether or not individual i attends college and v_i represents some unobserved characteristic that also affects individual i's income. The treatment effect is represented by b_i and this may vary across individuals.

We are interested in determining the treatment effect for each individual i.

$$b_i = y_i(1) - y_i(0) \qquad (4.2)$$

This is the difference between the two possible outcomes for each individual.

4.2.2   Simulation of Impossible Data

Imagine you have access to the impossibly good data set created below (actually, just an impossible data set). The data provides information on the simulated individual's outcome (y) for *both* treatments $(x = 0$ and $x = 1)$. This is equivalent to knowing an individual's income for both the case where they went to college and the case where they did not go to college. These counter-factual outcomes are called **potential outcomes** (Rubin, 1974).

---

[1] Alexandria Ocasio-Cortez is often referred to as AOC.

```
> set.seed(123456789)
> N <- 200
> a <- 2
> b <- rnorm(N,mean=2,sd=3)
> # this creates variation in the slope with an average
> # effect of 2.
> x0 <- rep(0,N) # creates a vector of zeros
> x1 <- rep(1,N)
> u <- rnorm(N)
> y <- a + b*cbind(x0,x1) + u
> # y is a matrix, [a + u, a + b + u]
> # rep creates a vector by repeating the first number by the
> # amount of the second number.
```

Figure 4.1 presents the density functions, means and cumulative distribution functions of the two potential outcomes for the simulated data. The figure suggests that individuals generally have better outcomes when $x = 1$. Let y be income and $x = 1$ be college attendance. Do you think this is evidence that people earn more money because they attend college? The mean of the distribution of income for those attending college is much higher than the mean of the distribution of income for those not attending college. Assuming that this simulated data represented real data, should AOC 2028 use these results as evidence for making college free?

A concern is that the two distributions overlap. Moreover, the cumulative distributions functions cross. There may be individuals in the data who are actually better off if $x = 0$. The average college attendee earns more than the average non-college attendee but some may earn less if they go to college. We can determine if this occurs by looking at the joint distribution of potential outcomes. We will see that the crossing observed in Figure 4.1 implies that some individuals are better off if $x = 0$ while others are better off if $x = 1$.

4.2.3 Distribution of the Treatment Effect

Equation (4.2) states that the treatment effect may vary across individuals. If it does, then it has a distribution. Figure 4.2 presents the density and cumulative distribution function for the difference in outcome if the individual attended college and if the individual did not. The distribution shows that the treatment effect varies across individuals. It is heterogeneous. Moreover, the effect of college may either increase or decrease income, depending on the individual.

```
> par(mfrow=c(2,1)) # creates a simple panel plot
> par(mar=c(2,4,0.5,0.5))  # adjusts margins between plots.
> plot(density(y[,1]),type="l",lwd=5,xlim=range(y),
+       ylab="density",main="")
> lines(density(y[,2]),lwd=2)
> abline(v=colMeans(y),lwd=c(5,2))
> legend("topright",c("No College","College"),lwd=c(5,2))
> plot(ecdf(y[,1]), xlim=range(y),main="",do.points=FALSE,
+       lwd=5,xlab="y")
> lines(ecdf(y[,2]),lwd=2,do.points=FALSE)
> # ecdf empirical cumulative distribution function.
```

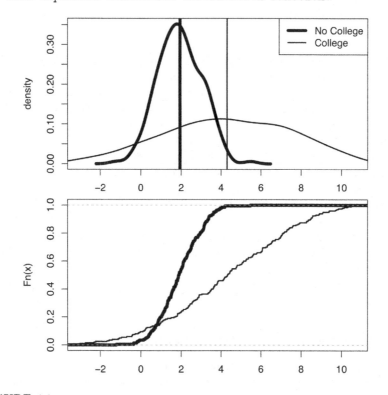

FIGURE 4.1

Density and cumulative probability distribution of potential outcomes. The
vertical lines are the mean potential outcomes.

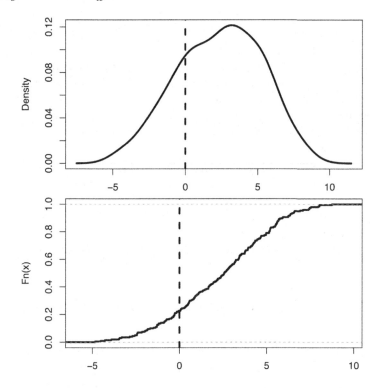

FIGURE 4.2

Density and cumulative probability distribution of the treatment effect. It is the distribution of the difference between income if the individual attends college and the income if they do not. The vertical lines are where the treatment effect is 0.

4.3 Average Treatment Effect

The **average treatment effect** (ATE) holds a special position in econometrics and statistics. A possible reason is that it measures the average difference in potential outcomes. That's actually pretty neat given that outside of our impossible data we cannot observe the difference in potential outcomes. How can we measure the average of something we cannot observe?

4.3.1 ATE and Its Derivation

```
> # mean of the difference vs difference of the means
> mean(y[,2]-y[,1]) == mean(y[,2]) - mean(y[,1])
```

[1] TRUE

The mean of the difference is equal to the difference of the means. We cannot observe the difference in the treatment outcomes. But, we can observe the outcomes of each treatment separately. We can observe the mean outcomes for each treatment. This neat bit of mathematics is possible because averages are linear operators.

We can write out the expected difference in potential outcomes by the Law of Total Expectations.

$$\mathbb{E}(Y_1 - Y_0) = \int_{y_0} \int_{y_1} (y_1 - y_0) f(y_1|y_0) f(y_0) dy_1 dy_0 \qquad (4.3)$$

where Y_x refers to the outcome that occurs if the individual receives treatment x. It is the potential outcome for x.

The rest follows from manipulating the conditional expectations.

$$
\begin{aligned}
\mathbb{E}(Y_1 - Y_0) &= \int_{y_0} \left(\int_{y_1} y_1 f(y_1|y_0) dy_1 - y_0 \right) f(y_0) dy_0 \\
&= \int_{y_0} \left(\int_{y_1} y_1 f(y_1|y_0) dy_1 \right) f(y_0) dy_0 - \int_{y_0} y_0 f(y_0) dy_0 \qquad (4.4) \\
&= \int_{y_1} y_1 f(y_1) dy_1 - \int_{y_0} y_0 f(y_0) dy_0 \\
&= \mathbb{E}(Y_1) - \mathbb{E}(Y_0)
\end{aligned}
$$

Rubin (1974) presents the derivation in Equation (4.4). He points out that if we can estimate each of the average potential outcomes then we have an estimate of the average treatment effect.

But can we estimate the average potential outcome?

4.3.2   ATE and Do Operators

To answer this question, it is clearer to switch notation. At the risk of upsetting the Gods of Statistics, I will mix notations from two different causal models. The expected potential outcome if $X = 1$ is assumed to be equal to the expected outcome conditional on $do(X) = 1$ (Pearl and Mackenzie, 2018).

$$\mathbb{E}(Y_1) = \mathbb{E}(Y|do(X) = 1) \qquad (4.5)$$

By "do" we mean that this is the expected outcome if individuals in the data faced a policy which forced the treatment $X = 1$. We are holding all other effects constant when the policy change is made. It is "do" as in "do a policy."

The notation highlights the fact that the expected potential outcome of a treatment may not be equal to expected outcomes in a particular treatment. In general, $\mathbb{E}(Y|do(X) = 1) \neq \mathbb{E}(Y|X = 1)$, where the second term is observed in the data. The second term is standard notation for the expected outcome among individuals observed in the data with the treatment equal to 1. This is do operator notation for "correlation does not imply causation."

To see why these numbers are not the same, consider the following derivation.

We can write down the expected outcome conditional on the do operator by the Law of Total Expectations. We can write out the average outcome conditional on the policy as the sum of the average outcomes of the policy conditional on the observed treatments weighted by the observed probabilities of the treatments.

$$
\begin{aligned}
\mathbb{E}(Y|\mathrm{do}(X)=1) \quad &= \mathbb{E}(Y|\mathrm{do}(X)=1, X=0)\Pr(X=0) \\
&+\mathbb{E}(Y|\mathrm{do}(X)=1, X=1)\Pr(X=1)
\end{aligned} \tag{4.6}
$$

The expected outcome under a policy in which individuals go to college is a weighted sum of the effect of the policy on individuals who currently go to college and the effect of the policy on individuals who currently do not go to college.

We are generally able to observe three of the four numbers on the right-hand side of Equation (4.6). We observe the probability individuals are allocated to the current treatments. In addition, we assume that $\mathbb{E}(Y|\mathrm{do}(X)=1, X=1) = \mathbb{E}(Y|X=1)$. That is, we assume that the expected outcome for people assigned to a treatment will be the same as if there was a policy that assigned them to the same treatment. The number we do not observe in the data is $\mathbb{E}(Y|\mathrm{do}(X)=1, X=0)$. We cannot observe the expected outcome conditional on a policy assigning a person to one treatment when they are observed receiving the other treatment. We cannot observe the expected income from attending college for people who do not attend college.

4.3.3 ATE and Unconfoundedness

We can estimate the average treatment effect if we are willing to make the following assumption.

Assumption 1. *Unconfoundedness.* $\mathbb{E}(Y|do(X) = x, X = x) = \mathbb{E}(Y|do(X) = x, X = x')$

Assumption 1 states that expected outcome of the policy does not vary with treatment observed in the data. Under the assumption, there is no information content in the fact that one group attends college and one group does not. This assumption may be reasonable if we have data from an ideal randomized controlled trial. For most other data, including many randomized controlled trials, the assumption may not be credible.

The assumption implies we can substitute the unknown expected value with the known expected value.

$$
\begin{aligned}
\mathbb{E}(Y|\mathrm{do}(X)=1) \quad &= \mathbb{E}(Y|\mathrm{do}(X)=1, X=0)\Pr(X=0) \\
&+\mathbb{E}(Y|\mathrm{do}(X)=1, X=1)\Pr(X=1) \\
&= \mathbb{E}(Y|\mathrm{do}(X)=1, X=1)\Pr(X=0) \\
&+\mathbb{E}(Y|\mathrm{do}(X)=1, X=1)\Pr(X=1) \\
&= \mathbb{E}(Y|X=1)
\end{aligned} \tag{4.7}
$$

The implication is that we can estimate the average of the potential outcomes for each treatment. Thus we can estimate the average difference in potential outcomes. Said differently, unconfoundedness allows us to estimate the average treatment effect.

4.3.4 ATE and Simulated Data

```
> X <- runif(N) < 0.3 # treatment assignment
> Y <- (1-X)*y[,1] + X*y[,2] # outcome conditional on treatment
```

Consider a change to our simulated data to make it look more like an actual data set. In the new data we only see one outcome and one treatment for each individual. However, if we can make the unconfoundedness assumption then we can estimate the average treatment effect. Our new data satisfies the assumption because the assignment to treatment is random.

```
> mean(Y[X==1]) - mean(Y[X==0])
```

```
[1] 2.432335
```

In the data the true average treatment effect is 2. Our estimate is 2.43. What changes could you make to the simulated data that would increase the accuracy of the estimate?[2]

4.4 Kolmogorov Bounds

There are policy questions where the ATE provides a useful answer, but it is often provided as a *statistic of convenience*. In the data generated above, many simulated individuals are better off under treatment $x = 1$. But not everyone is better off. It may be useful for policy makers to know something about the joint distribution of potential outcomes or the distribution of the treatment effect.[3]

We do not have access to the impossible data generated above. We cannot estimate the joint distribution of potential outcomes or the distribution of the treatment effect. However, we can **bound** these distributions.

4.4.1 Kolmogorov's Conjecture

The Russian mathematician, Andrey Kolmogorov, conjectured that difference of two random variables with known marginals could be bounded in the following way. Note that I have written this out in a simplified way that will look more closely like the way it is implemented in **R**.[4]

---

[2]Some of these changes are discussed in Chapter 1.
[3]This chapter discusses the second, but the two are mathematically related.
[4]See Fan and Park (2010) and the citations therein.

Theorem 1. *Kolmogorov's Conjecture. Let $\beta_i = y_i(1) - y_i(0)$ denote the treatment effect and F denote its distribution. Let F_0 denote the distribution of outcomes for treatment $(x = 0)$ and F_1 denote the distribution of outcomes for treatment $(x = 1)$. Then $F^L(b) \leq F(b) \leq F^U(b)$, where*

$$F^L(b) = \max\{\max_y F_1(y) - F_0(y - b), 0\} \qquad (4.8)$$

and

$$F^U(b) = 1 + \min\{\min_y F_1(y) - F_0(y - b), 0\} \qquad (4.9)$$

Theorem 1 states that we can bound the distribution of the treatment effect even though we only observe the distributions of outcomes for each of the treatments. You may be surprised to learn how easy these bounds are to implement and how much information they provide about the distribution of the treatment effect.

4.4.2 Kolmogorov Bounds in R

We can use Theorem 1 as pseudo-code for the functions that bound the treatment effect distribution.

```
> FL <- function(b, y1, y0) {
+     f <- function(x) -(mean(y1 < x) - mean(y0 < x - b))
+     # note the negative sign as we are maximizing
+     # (Remember to put it back!)
+     a <- optimize(f, c(min(y1,y0),max(y1,y0)))
+     return(max(-a$objective,0))
+ }
> FU <- function(b, y1, y0) {
+     f <- function(x) mean(y1 < x) - mean(y0 < x - b)
+     a <- optimize(f, c(min(y1,y0), max(y1,y0)))
+     return(1 + min(a$objective,0))
+ }
```

Figure 4.3 presents the distribution of the treatment effect for the simulated data as well as the lower and upper bounds. Remember that in normal data we cannot observe the treatment effect but thanks to math we can determine its bounds. If you look closely, you will notice that some simulated individuals must be harmed by the treatment. At 0, the bounds are strictly positive. Of course, we know that in our impossible data some simulated individuals are in fact worse off.

```
> K <- 50
> min_diff <- min(y[,1]) - max(y[,2])
> max_diff <- max(y[,1]) - min(y[,2])
> delta_diff <- (max_diff - min_diff)/K
> y_K <- min_diff + c(1:K)*delta_diff
> plot(ecdf(y[,2] - y[,1]), do.points=FALSE,lwd=3,main="")
> lines(y_K,sapply(y_K, function(x) FL(x,y[,2],y[,1])),
+       lty=2,lwd=3)
> lines(y_K,sapply(y_K, function(x) FU(x,y[,2],y[,1])),
+       lty=3,lwd=3)
> abline(v=0,lty=2,lwd=3)
```

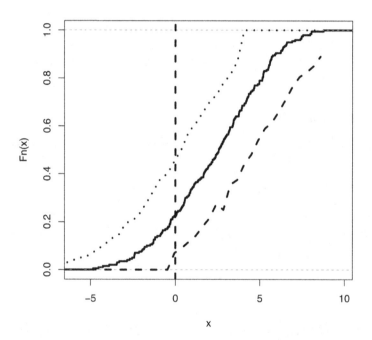

FIGURE 4.3
Distribution of the treatment effect along with its upper and lower bounds.

4.5 Do "Nudges" Increase Savings?

Researchers in economics and psychology have found that individuals often make poor decisions. They make decisions that are against the individual's own interest. Given this, can policies or products be provided that "nudge" individuals to make better decisions?

Ashraf et al. (2006) describe an experiment conducted with a bank in the Philippines. In the experiment some customers were offered "commitment" savings accounts. In these accounts the customer decides upon a goal, such as a target amount or a target date, and can deposit but not withdraw until the goal is reached. Such products may help people with issues controlling personal finances or interacting with household members on financial matters. People offered accounts did not actually have to open an account and many did not.

Ashraf et al. (2006) use a **field experiment** to determine the effectiveness of a commitment savings account.[5] In the experiment there are three treatment groups; the first group is offered the commitment savings account at no extra cost or savings, the second group is provided information on the value of savings, and the third is a control. Here we will compare the commitment group to the control.

The section uses the data to illustrate the value of Kolmogorov bounds.

4.5.1   Field Experiment Data

We first replicate the findings in Ashraf et al. (2006). The data is available at https://doi.org/10.7910/DVN/27854 or at https://sites.google.com/view/microeconometricswithr/table-of-contents.[6]

```
> require(readstata13)
> # this data set was saved with Stata version 13.
> x <- read.dta13("seedanalysis_011204_080404.dta")
> index_na <- is.na(rowSums(cbind(x$treatment,
+                                 x$balchange,x$marketing)))==0
> x1 <- x[index_na,]
> bal_0 <- x1[x1$treatment==0 & x1$marketing==0,]$balchange
> bal_1 <- x1[x1$treatment==1 & x1$marketing==0,]$balchange
> # we are just going to look at the people who did not receive
> # the marketing information.
> # These people are split between those that received
> # the account
> # (treatment = 1), and those that did not (treatment = 0).
> # balchange - measure their balance changed in a year.
> lbal_0 <- log(bal_0 + 2169)
> lbal_1 <- log(bal_1 + 2169)
> # the distribution of balances is very skewed.
> mean(bal_1) - mean(bal_0)

[1] 411.4664
```

The average treatment effect is a 411 peso increase (about \$200) in savings

---

[5]Field experiments are randomized trials in which people or villages or schools are assigned between trial arms.

[6]Note that the version on the original website has a slightly different name.

after 12 months for those offered the commitment accounts. This result suggests that commitment accounts have a significant impact on savings rates. However, it is not clear if everyone benefits and how much benefit these accounts provide.

4.5.2   Bounds on the Distribution of Balance Changes

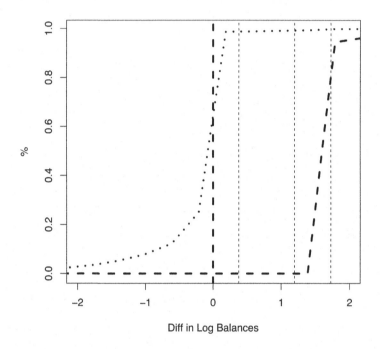

FIGURE 4.4
Upper and lower bounds on the distribution of difference in log balances between the treatment and the control. Marks at 1,000, 5,000 and 10,000 using the transformation above.

Figure 4.4 presents the bounds on the distribution of the treatment effect. The figure shows that there is a small portion of the population that ends up saving a large amount due to the commitment savings device, over 10,000 pesos. It also shows that for a large part of the population the commitment savings may or *may not* increase savings. There may even be people who actually end up saving less. Unlike the example above, we cannot show that the fraction must be greater than 0.

4.5.3 Intent To Treat Discussion

One issue with the analysis presented above, and with the main results of Ashraf et al. (2006), is that they are the **intent to treat** estimates. We have estimated the treatment effect of being "assigned" to a commitment account. People are not lab rats. They have free will. In this case, people assigned to the commitment accounts had the choice of whether to open the account or not. Many did not.

Can you calculate the average treatment effect using the instrumental variable approach? Hint: it is much higher. Did you calculate the ATE or the LATE?

More generally, the concern is that we do not know what would happen to the savings of people who were assigned the commitment account but chose not to open it. Did these people know something that we do not?

4.6 Manski Bounds

In his seminal paper, *Non-Parametric Bounds on Treatment Effects*, Chuck Manski introduced the idea of set estimation to economics (Manski, 1990). Manski argues that many of the assumptions underlying standard econometrics are *ad hoc* and unjustified. Rather than making such assumptions, Manski suggests presenting results based on assumptions that can be well justified. In many cases, such assumptions do not provide precise estimates.

Manski also points out that the econometrician and the policy maker may have different views on the reasonableness of assumptions. Therefore, the econometrician should present the results ordered from those based on the most reasonable assumptions to those results based on the least reasonable assumptions. This approach to presenting research gives the policy maker a better understanding of the relationship between the assumptions and the policy predictions (Manski and Pepper, 2013).

The section presents the bounds approach and illustrates it with simulated data.

4.6.1 Confounded Model

Consider the following confounded version of the model presented above.

$$y_i(x_i) = a + b_i x_i + v_{1i} \tag{4.10}$$

where y_i is individual i's income, $x_i \in \{0, 1\}$ is whether or not individual i attends college and v_{1i} represents some unobserved characteristic. The treatment effect is represented by b_i and this may vary across individuals.

This time, the value of the policy variable is also determined by the unobserved characteristic that determines income.

$$x_i^* = f + cv_{1i} + dz_i + v_{2i}$$

$$x_i = \begin{cases} 1 & \text{if } x_i^* > 0 \\ 0 & \text{otherwise} \end{cases} \tag{4.11}$$

where x_i^* is a latent (hidden) variable that determines whether or not the individual attends college. If the value of the latent value is high enough, then the individual attends college. Importantly, the value of this latent variable is determined by the same unobserved characteristic that determines income. That is, if v_{1i} is large and the parameter c is positive, then y_i will tend to be larger when x_i is 1 and lower when x_i is 0.

4.6.2   Simulation of Manski Bounds

Consider simulated data from a **confounded** data set.

```
> c <- 2
> d <- 4
> f <- -1
> Z <- round(runif(N))
> u_2 <- rnorm(N)
> x_star <- f + c*u + d*Z + u_2
> X <- x_star > 0 # treatment assignment
> Y <- (1-X)*y[,1] + X*y[,2] # outcome conditional on treatment
> mean(Y[X==1]) - mean(Y[X==0])
```

```
[1] 3.506577
```

The simulated data illustrates the problem. If we assume **unconfoundedness**, we can estimate the average treatment effect. Our estimate is not close to the true value of 2. Try running OLS of y on x. What do you get?

In economics we call this a **selection problem**. One solution is to use an instrumental variable estimate to determine b. But what if we don't have an instrument? What if we don't believe the assumptions of the IV model are credible given our data? An alternative to making an unreasonable assumption is to bound the value of interest.

4.6.3   Bounding the Average Treatment Effect

The average treatment effect of college is the difference in the expected outcome given a policy of going to college and a policy of not going to college.

$$ATE = \mathbb{E}(Y|\text{do}(X) = 1) - \mathbb{E}(Y|\text{do}(X) = 0) \tag{4.12}$$

From above we know it can be written as the difference in expected income when the policy forces everyone to go to college and the expected income when the policy forces everyone not to go to college.

We can write out this via the Law of Total Expectation.

$$
\begin{aligned}
ATE \ &= \mathbb{E}(Y|\mathrm{do}(X) = 1, X = 1)\Pr(X = 1)+ \\
&\quad \mathbb{E}(Y|\mathrm{do}(X) = 1, X = 0)\Pr(X = 0) \\
&\quad - (\mathbb{E}(Y|\mathrm{do}(X) = 0, X = 1)\Pr(X = 1)+ \\
&\quad \mathbb{E}(Y|\mathrm{do}(X) = 0, X = 0)\Pr(X = 0))
\end{aligned}
\tag{4.13}
$$

Each expectation can be split into the group that attends college and the group that does not attend college. We observe the outcome of the policy that sends the individuals to college for the group that actually goes to college. If we assume that their outcome from the policy is the same as we observe, then we can substitute the observed values into the equation.

$$
\begin{aligned}
ATE \ &= \Pr(X = 1)\left(\mathbb{E}(Y|X = 1) - \mathbb{E}(Y|\mathrm{do}(X) = 0, X = 1)\right) \\
&\quad + \Pr(X = 0)\left(\mathbb{E}(Y|\mathrm{do}(X) = 1, X = 0) - \mathbb{E}(Y|X = 0)\right)
\end{aligned}
\tag{4.14}
$$

We don't know the outcome of the policy that sends individuals to college for the group that actually does not go to college. Note that I rearranged the equation a little.

We cannot determine the ATE. But we can **bound** the ATE by replacing the values we cannot observe with values we can observe. Importantly, we *know* these observed values must be larger (smaller) than the values we cannot observe.

4.6.4 Natural Bounds of the Average Treatment Effect

What is the weakest assumption we could make? An expectation is bounded by the smallest possible value and the largest possible value. An average cannot be smaller than the smallest possible value in the set being averaged. Similarly, the average cannot be larger than the largest possible value in the set being averaged.

The bounds are created by replacing the unknown values with the smallest (largest) values they could be. Let $\underline{Y}$ represent the lower bound (the lowest possible value) and $\overline{Y}$ represent the upper bound (the largest possible value). Manski calls this the **worst-case bounds**, while Pearl uses the term **natural bounds**.[7]

Given these values, we can calculate the bounds on the average treatment effect.

$$
\overline{ATE} = (\mathbb{E}(Y|X = 1) - \underline{Y})\Pr(X = 1) + (\overline{Y} - \mathbb{E}(Y|X = 0))\Pr(X = 0)
$$

$$
\underline{ATE} = (\mathbb{E}(Y|X = 1) - \overline{Y})\Pr(X = 1) + (\underline{Y} - \mathbb{E}(Y|X = 0))\Pr(X = 0)
\tag{4.15}
$$

---

[7]If we don't know the possible values, we could use the observed values. This assumption may be less "natural" than we may prefer.

Note how the bounds on the ATE are calculated. The maximum on the ATE is denoted by the overline. It is when the first expected outcome is as large as possible and the second expected outcome is as small as possible. Similarly, the minimum on the ATE is when the first outcome is as small as possible and the second outcome is as large as possible. The minimum on the ATE is denoted by the underline.

4.6.5   Natural Bounds with Simulated Data

In the simulated data we can use the observed minimum and maximum.

```
> PX1 = mean(X==1)
> PX0 = mean(X==0)
> EY_X1 = mean(Y[X==1])
> EY_X0 = mean(Y[X==0])
> minY = min(Y)
> maxY = max(Y)
```

The bounds are calculated by replacing the unknown outcome with the minimum possible value of the outcome and, alternatively, the maximum possible value for the outcome.

```
> # ATE upper bound
> (EY_X1 - minY)*PX1 + (maxY - EY_X0)*PX0

[1] 7.975223

> # ATE lower bound
> (EY_X1 - maxY)*PX1 + (minY - EY_X0)*PX0

[1] -5.010368
```

These bounds are wide. The average treatment effect of X on Y is between -5.01 and 7.98. The true value is 2.

4.6.6   Are Natural Bounds Useless?

The bounds presented above are wide and don't even predict the correct sign for the ATE. What can we take away from this information?

First, if we are unwilling to make stronger assumptions, then the data may simply not help us answer the policy question of interest. Manski calls the willingness to make incredible assumptions in order to get more certain results, the "lure of incredible certitude" (Manski, 2020). He argues that this practice reduces the public and the policy maker's willingness to rely on science and accept new knowledge.

Second, it is not that we don't learn anything from the data. In this case we learn that the effect of a policy $do(X) = 1$ cannot have a larger effect than 8. There are cases where this information may be enough for policy makers to seek an alternative. For example, a cost benefit analysis may have suggested that for a policy to be of value, the effect of the policy must be greater in magnitude than 8. In that case, the bounds provide enough information to say that the policy's benefits are outweighed by its costs.

Third, there may be assumptions and data that are reasonable and allow tighter bounds. Those are discussed more in the following sections.

4.6.7 Bounds with Exogenous Variation

We may have **tighter** bounds through variation in the data. In particular, we need variation such that the effect of the policy doesn't change across different subsets of the data, but the bounds do.

Assumption 2.

$$\mathbb{E}(Y|do(X) = 1, Z = z) - \mathbb{E}(Y|do(X) = 0, Z = z)$$
$$= \mathbb{E}(Y|do(X) = 1, Z = z') - \mathbb{E}(Y|do(X) = 0, Z = z') \quad (4.16)$$

for all z, z'.

Assumption 2 is like an instrumental variables assumption. Manski calls it a **level-set** assumption.[8] It states that there exists some observable characteristic such that the average treatment effect does not change with changes in the observable characteristic. Given this property it is possible to get **tighter** bounds by estimating the bounds on the average treatment effect for various subsets of the data. Under the assumption, the average treatment effect must lie in the intersection of these bounds. Thus the new bounds are the intersection of these estimated bounds.

$$
\begin{aligned}
\overline{ATE} &= \min\{(\mathbb{E}(Y|X = 1, Z = 1) - \underline{Y})\Pr(X = 1|Z = 1) \\
&\quad + (\overline{Y} - \mathbb{E}(Y|X = 0, Z = 1))\Pr(X = 0|Z = 1), \\
&\quad (\mathbb{E}(Y|X = 1, Z = 0) - \underline{Y})\Pr(X = 1|Z = 0) \\
&\quad + (\overline{Y} - \mathbb{E}(Y|X = 0, Z = 0))\Pr(X = 0|Z = 0)\} \\
\\
\underline{ATE} &= \max\{(\mathbb{E}(Y|X = 1, Z = 1) - \overline{Y})\Pr(X = 1|Z = 1) \\
&\quad + (\underline{Y} - \mathbb{E}(Y|X = 0, Z = 1))\Pr(X = 0|Z = 1), \\
&\quad (\mathbb{E}(Y|X = 1, Z = 0) - \overline{Y})\Pr(X = 1|Z = 0) \\
&\quad + (\underline{Y} - \mathbb{E}(Y|X = 0, Z = 0))\Pr(X = 0|Z = 0)\}
\end{aligned}
$$

$$(4.17)$$

These are the bounds when the instrument-like variables has two values $(Z \in \{0, 1\})$.

---

[8]Would an IV estimator discussed in Chapter 3 satisfy Assumption 2?

4.6.8   Exogenous Variation in Simulated Data

We haven't used it yet, but there is a variable Z in the simulated data that
is associated with changes in the policy variable but does not directly affect
income.[9]

```
> EY_X1Z1 = mean(Y[X==1 & Z==1])
> EY_X1Z0 = mean(Y[X==1 & Z==0])
> EY_X0Z1 = mean(Y[X==0 & Z==1])
> EY_X0Z0 = mean(Y[X==0 & Z==0])
> PX1_Z1 = mean(X[Z==1]==1)
> PX1_Z0 = mean(X[Z==0]==1)
> PX0_Z1 = mean(X[Z==1]==0)
> PX0_Z0 = mean(X[Z==0]==0)

> # ATE upper bound
> min((EY_X1Z1 - minY)*PX1_Z1 + (maxY - EY_X0Z1)*PX0_Z1,
+       (EY_X1Z0 - minY)*PX1_Z0 + (maxY - EY_X0Z0)*PX0_Z0)

[1] 7.049019

> # ATE lower bound
> max((EY_X1Z1 - maxY)*PX1_Z1 + (minY - EY_X0Z1)*PX0_Z1,
+       (EY_X1Z0 - maxY)*PX1_Z0 + (minY - EY_X0Z0)*PX0_Z0)

[1] -4.00698
```

We see that using the level-set restriction we do get **tighter** bounds, but
the change is not very large. What changes could you make in the simulated
data to get a larger effect of using the **level-set restriction**?

4.6.9   Bounds with Monotonicity

Can the bounds be tighter with some economics? Remember that we observe
the cases where $do(X) = x$ and $X = x$ match. We don't observe the cases
where they don't match. However, we can use the observed cases to bound
the unobserved cases. Mathematically, there are a couple of options regarding
which observed outcomes can be used for the bounds. Which option you choose
depends on the economics.

In the simulated data a higher unobserved term is associated with a greater
likelihood of choosing treatment $x = 1$. That is, holding everything else constant,
observing someone receiving treatment $x = 1$ means that they will have higher
outcomes. This is a **monotonicity** assumption. In math, the assumption is as
follows.

Assumption 3. *Monotonicity.* $\mathbb{E}(Y|do(X) = 1, X = 1) \geq \mathbb{E}(Y|do(X) = 1, X = 0)$ *and* $\mathbb{E}(Y|do(X) = 0, X = 0) \leq \mathbb{E}(Y|do(X) = 0, X = 1)$

---

[9]Does Z satisfy the assumptions of an instrumental variable?

Assumption 3 states that observing someone receive treatment $x = 1$ tells us about their unobserved term. For example, if we hold the treatment the same for everyone, then the people who choose $x = 1$ will have higher expected outcomes. Those that are "selected" into college may have better returns to schooling than the average person. The treatment has monotonic effects on outcomes. We can use this assumption to tighten the bounds on the ATE. In particular, the upper bound can be adjusted down.

$$\overline{\mathbb{E}(Y|\mathrm{do}(X) = 1)} = \mathbb{E}(Y|X = 1)$$

$$\underline{\mathbb{E}(Y|\mathrm{do}(X) = 0)} = \mathbb{E}(Y|X = 0)$$

(4.18)

The monotonicity assumption implies that forcing everyone into treatment $x = 1$ cannot lead to better expected outcomes than the outcomes we observe given the treatment. Similarly, forcing everyone into treatment $x = 0$ cannot have a worse expected outcome than the outcomes we observe given the treatment.

$$\overline{ATE} = (\overline{Y} - \mathbb{E}(Y|X = 0))\Pr(X = 0)$$

$$\underline{ATE} = (\mathbb{E}(Y|X = 1) - \overline{Y})\Pr(X = 1)$$

(4.19)

4.6.10 Bounds with Monotonicity in Simulated Data

```
> # ATE upper bound
> (maxY - EY_X0)*PX0

[1] 3.76668

> # ATE lower bound
> (EY_X1 - maxY)*PX1

[1] -3.64774
```

Imposing Assumption 3 on the simulated data allows us to **tighten** the bounds. They reduce down to $[-3.65, 3.77]$. Remember the true average in the simulated data is 2. It lowers the potential value of the treatment from 8 to 4.

Note that the impact of these assumptions is presented in the order that Manski and Pepper (2013) prefer. We started with the most credible assumption, the **natural bounds**. Then we moved to make a level-set restriction because we had a variable that satisfied the assumption. Finally, we made the monotonicity assumption.

4.7   More Guns, Less Crime?

One of the most controversial areas in microeconometrics is estimating the effect of various gun laws on crime and gun related deaths. To study these effects, economists and social scientists look at how these laws vary across the United States and how those changes in laws are related to changes in crime statistics (Manski and Pepper, 2018).

Justice Louis Brandeis said that a "state may, if its citizens choose, serve as a laboratory; and try novel social and economic experiments without risk to the rest of the country."[10] The US states are a "laboratory of democracy." As such, we can potentially use variation in state laws to estimate the effects of those laws. The problem is that US states are very different from each other. In the current terminology, the states with strong pro-gun laws tend to be "red" states or at least "purple" states. They also tend to have large rural populations.

Between 1980 and 1990, twelve states adopted right to carry (RTC) laws. We are interested in seeing how crime fared in those states relative to states that did not adopt those laws. To do this we can look at crime rates from the 1980s and 1990s. A potential problem is that the crack epidemic hit the United States at exactly this time, rising through the 80s and 90s before tailing off. The crack cocaine epidemic was associated with large increases in crime rates in urban areas (Aneja et al., 2011).

This section uses publicly available crime data to illustrate the value of the bounds approach.

4.7.1   Crime Data

The data is downloaded from John Donohue's website.[11] While there is quite a lot of variation in gun laws, the definition of RTC is "shall issue" in data set used. For crime, we use the per population rate of aggravated assaults per state, averaged over the post 1990 years. The code also calculates the physical size of the state, which is a variable that will be used later.

```
> library(foreign)
> # the data is standard Stata format, the library foreign
> # allows this data to be imported.
> x <- read.dta("UpdatedStateLevelData-2010.dta")
> Y <- X <- Z <- NULL
> # the loop will create variables by adding to the vectors
> for (i in 2:length(unique(x$state))) {
```

---

[10]See New State Ice Co vs. Liebmann 285 US 262 (1932).

[11]https://works.bepress.com/john_donohue/89/ or https://sites.google.com/view/microeconometricswithr/table-of-contents

```
+    # length measures the number of elements in the object.
+    state = sort(unique(x$state))[i]
+    # note the first state is "NA"
+    X <- c(X,sum(x[x$state==state,]$shalll, na.rm = TRUE) > 0)
+    # determines if a state has an RTC law at
+    # some point in time.
+    # na.rm tells the function to ignore NAs
+    Y <- c(Y,mean(x[x$state==state & x$year > 1990,]$rataga,
+                      na.rm = TRUE))
+    # determines the average rate of aggrevated assualt for the
+    # state post 1990.
+    Z <- c(Z,mean(x[x$state==state & x$year > 1990,]$area,
+                      na.rm = TRUE) > 53960)
+    # determines the physical area of the state
+    # Small state = 0, large stage = 1
+    # print(i)
+ }
```

Figure 4.5 shows the histogram for the average aggravated assault rate per state in the post 1990 years. It shows that rate per 100,000 is between 0 and 600 for the most part.

4.7.2   ATE of RTC Laws under Unconfoundedness

If we assume unconfoundedness, then RTC laws lower aggravated assault. Comparing the average rate of aggravated assault in states with RTC laws to states without RTC laws, we see that the average is lower with RTC laws.

```
> EY_X1 <- mean(Y[X==1])
> EY_X0 <- mean(Y[X==0])
> EY_X1 - EY_X0
```

```
[1] -80.65852
```

Unconfoundedness is not a reasonable assumption. We are interested in estimating the average effect of implementing an RTC law. We are not interested in the average rate of assaults conditional on the state having an RTC law.

4.7.3   Natural Bounds on ATE of RTC Laws

We cannot observe the effect of RTC laws for states that do not have RTC laws. We could assume that the assault rate lies between 0 and 100,000 (which it does).

```
> PX0 <- mean(X==0)
> PX1 <- mean(X==1)
```

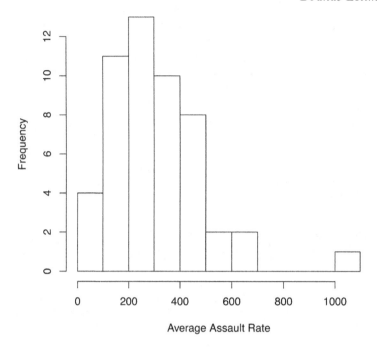

FIGURE 4.5
Histogram of average aggregated assault per 100,000 by state.

```
> minY <- 0
> maxY <- 100000
> # ATE upper bound
> (EY_X1 - minY)*PX1 + (maxY - EY_X0)*PX0

[1] 23666.01

> # ATE lower bound
> (EY_X1 - maxY)*PX1 + (minY - EY_X0)*PX0

[1] -76333.99
```

The natural bounds are very very wide. An RTC policy may lead to assault rates decreasing by 75,000 or increasing by 24,000 per 100,000 people.

We can make these bounds tighter by assuming that assault rates of the policy cannot lie outside the rates observed in the data.

```
> minY <- min(Y)
> maxY <- max(Y)
> # ATE upper bound
> (EY_X1 - minY)*PX1 + (maxY - EY_X0)*PX0
```

```
[1]  334.1969

> # ATE lower bound
> (EY_X1 - maxY)*PX1 + (minY - EY_X0)*PX0

[1]  -624.7655
```

These bounds are lot tighter. A policy that introduces RTC for the average state could decrease the assault rate by 625 or increase the assault rate by 334. Given that range, it could be that RTC laws substantially reduce aggravated assaults or it could be that they have little or no effect. They may even cause an increase in aggravated assaults.

4.7.4   Bounds on ATE of RTC Laws with Exogenous Variation

```
> PX1_Z1 <- mean(X[Z==1]==1)
> PX1_Z0 <- mean(X[Z==0]==1)
> PX0_Z1 <- mean(X[Z==1]==0)
> PX0_Z0 <- mean(X[Z==0]==0)
> EY_X1Z1 <- mean(Y[X==1 & Z==1])
> EY_X1Z0 <- mean(Y[X==1 & Z==0])
> EY_X0Z1 <- mean(Y[X==0 & Z==1])
> EY_X0Z0 <- mean(Y[X==0 & Z==0])
> # check your code here.
> # a NaN error below maybe due to a typo above.
> # NaN error may occur because the vector is all NAs
> # ATE upper bound
> min((EY_X1Z1 - minY)*PX1_Z1 + (maxY - EY_X0Z1)*PX0_Z1,
+      (EY_X1Z0 - minY)*PX1_Z0 + (maxY - EY_X0Z0)*PX0_Z0)

[1]  323.2504

> # ATE lower bound
> max((EY_X1Z1 - maxY)*PX1_Z1 + (minY - EY_X0Z1)*PX0_Z1,
+      (EY_X1Z0 - maxY)*PX1_Z0 + (minY - EY_X0Z0)*PX0_Z0)

[1]  -613.3812
```

We can make a level-set assumption. Assume that the instrument-like variable is the physical size of the state. The assumption is that the average treatment effect of implementing an RTC law must be the same irrespective of the physical size of the state. Note that observable outcomes like the assault rate and the proportion of states with RTC laws may vary with the physical size. The assumption is on the average treatment effect which is unobserved.

The bounds are tighter, although not much. RTC laws could reduce aggravated assaults by 613 or increase rates by 323.

4.7.5   Bounds on ATE of RTC Laws with Monotonicity

Would it be reasonable to use the monotonicity assumption above (Assumption 3)?

Let's assume that states that currently have RTC laws will also tend to have lower levels of aggravated assault. Moreover, forcing states that don't currently have RTC laws will not reduce the expected aggravated assaults below that level. This is the "negative" of the monotonicity assumption in the simulated data.

We can summarize this with Assumption 4.

Assumption 4. $\mathbb{E}(Y|do(X) = 1, X = 1) \leq \mathbb{E}(Y|do(X) = 1, X = 0)$ *and* $\mathbb{E}(Y|do(X) = 0, X = 0) \geq \mathbb{E}(Y|do(X) = 0, X = 1)$

Assumption 4 implies the following change to the bounds on the unobserved expectations.

$$\underline{\mathbb{E}(Y|\mathrm{do}(X) = 1)} = \mathbb{E}(Y|X = 1)$$

$$\overline{\mathbb{E}(Y|\mathrm{do}(X) = 0)} = \mathbb{E}(Y|X = 0)$$

(4.20)

Plugging these into the bounds on the ATE we have the following bounds on the effect of the RTC laws.

```
> # ATE upper bound
> (EY_X1 - minY)*PX1

[1]  184.2203

> # ATE lower bound
> (minY - EY_X0)*PX0

[1]  -75.66166
```

These bounds are substantially **tighter**. They suggest that the estimate of the ATE under unconfoundedness is actually at the high end of the possible effect of RTC laws. This is evidence that the **unconfoundedness** assumption cannot hold. At least, it is inconsistent with the weaker monotonicity assumption.

The results in this section suggest the slogan may be more accurately stated as "more guns, more or less crime."

4.8   Discussion and Further Reading

This chapter argues that it may be better to provide less precise estimates than precise predictions of little value to policy makers.

I strongly believe that the average treatment effect is given way too much prominence in economics and econometrics. ATE can be informative, but it can also badly mislead policy makers and decision makers. If we know the joint distribution of potential outcomes, then we may be able to better calibrate the policy. I hope that Kolmogorov bounds will become a part of the modern econometrician's toolkit. A good place to learn more about this approach is Fan and Park (2010). Mullahy (2018) explores this approach in the context of health outcomes.

Chuck Manski revolutionized econometrics with the introduction of set identification. He probably does not think so, but Chuck has changed the way many economists and most econometricians think about problems. We think much harder about the assumptions we are making. Are the assumptions credible? We are much more willing to present bounds on estimates, rather than make non-credible assumptions to get point estimates.

Manski's natural bounds allow the researcher to estimate the potential effect of the policy with minimal assumptions. These bounds may not be informative, but that in and of itself is informative. Stronger assumptions may lead to more informative results but at the risk that the assumptions, not the data, determine the results.

I highly recommend any book by Chuck Manski. However, Manski (1995) is the standard on non-parametric bounds. To understand more about potential outcomes see Rubin (1974). To understand more about **do operators** see Pearl and Mackenzie (2018).

Manski and Pepper (2018) use the bounds approach to analyze the relationship between right to carry laws and crime.

Part II

Structural Estimation

5

Estimating Demand

5.1 Introduction

In the early 1970s, San Francisco was completing a huge new infrastructure project, the Bay Area Rapid Transit (BART) system. The project initially cost $1.6 billion and included tunneling under the San Francisco Bay. Policy makers were obviously interested in determining how many people would use the new system once it was built. But that is a problem. How do you predict the demand for a product that does not exist?

One solution is to ask people. A survey was conducted of people who were likely to use the new transport system. The survey asked detailed questions about their current mode of transport and asked them whether they would use the new system. The concern is that it is hard for people to predict how they would use something that does not exist. Berkeley econometrician, Dan McFadden, suggested an alternative approach. Instead of asking people to predict what they would do, McFadden suggested using information on what people actually do do, then use economic theory to predict what they would do.

McFadden argued that combining survey data with economic theory would produce more accurate estimates than the survey data alone (McFadden, 1974). In the case of the BART survey, McFadden was correct. According to the survey data, 15% of respondents said that they would use BART. McFadden estimated that 6% of respondents would use BART. In fact, 6% of respondents actually did use BART.[1] Survey data is valuable, but people give more accurate answers to some questions than others.

The first part of the book discussed how **exogenous** variation is needed to use observed data to predict policy outcomes. Chapters 1 and 2 assume that observed variation in exposure to a policy is determined independently of unobserved characteristics. Chapters 3 and 4 relaxed this assumption but allowed economic theory to be used in estimating the impact of the policy. This part of the book extends the idea of using economic theory. This chapter introduces the idea of using **revealed preference**.

Today, the ideas that McFadden developed for analyzing the value of BART are used across economics, antitrust, marketing, statistics and machine

---

[1] https://www.nobelprize.org/prizes/economic-sciences/2000/mcfadden/lecture/

learning. At the Federal Trade Commission and the Department of Justice, economists use these techniques to determine whether a merger between ice cream manufacturers, or cigarette manufacturers, or supermarkets, or hospitals, will lead to higher prices.

When Google changed the way it displayed search results, user traffic moved away from Google's competitors. Such actions by a dominant firm like Google could lead to antitrust actions unless the changes also made users better off. By combining economic theory and data on the behavior of Google's users, we can determine whether Google's changes were pro or anti-competitive. According to the FTC's statement on the Google investigation, analysis of Google's click through data by staff economists showed that consumers benefited from the changes that Google made. This, and other evidence, led the FTC to end its investigation of Google's "search bias" practice with a 5-0 vote.[2]

The chapter begins with basic economic assumption of demand estimation, **revealed preference**. It takes a detour to discuss the **maximum likelihood algorithm**. It returns with Daniel McFadden's model of demand. The chapter introduces the **logit** and **probit** estimators, and uses them to determine whether consumers in small US cities value rail as much as their big city neighbors.

5.2 Revealed Preference

McFadden's analysis, and demand analysis more generally, relies on the following assumption.

Assumption 5. *Revealed Preference. If there are two choices, A and B, and we observe a person choose A, then her utility from A is greater than her utility from B.*

Assumption 5 states that if we observe someone choose product A when product B was available, then that someone prefers product A to product B. The assumption allows the researcher to infer unobserved characteristics from observed actions. It is also a fundamental assumption of microeconomics. Do you think it is a reasonable assumption to make about economic behavior?

The section uses simulated data to illustrate how revealed preference is used to estimate consumer preferences.

5.2.1 Modeling Demand

Consider a data set where a large number of individuals are observed purchasing either product A or product B at various prices for the two products. Each

---

[2]https://www.ftc.gov/system/files/documents/public_statements/295971/130103googlesearchstmtofcomm.pdf

individual will have some unobserved characteristic u, that we can call **utility**. We make two important assumptions about the individuals. First, their value for purchasing one of the two products is $u_{Ai} - p_A$. It is equal to their unobserved utility for product A minus the price of product A. Their utility is **linear in money**.[3] Second, if we observe person i purchase A at price p_A then we know that the following inequality holds.

$$u_{Ai} - p_A > u_{Bi} - p_B$$
or $\qquad\qquad$ (5.1)
$$u_{Ai} - u_{Bi} > p_A - p_B$$

Person i purchases good A if and only if her relative utility for A is greater than the relative price of A.

We usually make a transformation to **normalize** everything relative to one of the available products. That is, all prices and demand are made relative to one of the available products. Here we will **normalize** to product B. So $p = p_A - p_B$ and $u_i = u_{Ai} - u_{Bi}$. Prices and utility are net of prices and utility for product B.

In addition, we often observe the data at the market level rather than the individual level. That is, we see the fraction of individuals that purchase A. Below this fraction is denoted s. It is the share of individuals that purchase A.

5.2.2 Simulating Demand

The simulated data illustrates the power of the revealed preference assumption. Consider the following distribution of an unobserved term. The unobserved term is drawn from a normal distribution with a mean of 1 and a variance of 9 ($u \sim \mathcal{N}(1, 9)$). Let's assume that we have data on 1,000 people and each of them is described by this u characteristic.

```
> set.seed(123456789)
> N <- 1000
> u <- sort(rnorm(N, mean=1, sd=3))
```

Can we uncover this distribution from observed behavior of the individuals in our simulated data? Can we use the **revealed preference** assumption to uncover the unobserved term (u)?

```
> p <- 2
> mean(u - p > 0) # estimated probability

[1] 0.386

> 1 - pnorm(p, mean=1, sd=3) # true probability
```

---

[3] This is a standard assumption in economics. It means that the unobserved **utility** can be thought of as money.

```
[1] 0.3694413
```

If $p = 2$, then the share of people who purchase A is 39%, which is approximately equal to the probability that u is greater than 2. Combining the revealed preference assumption with the observed data allows us to uncover the fraction of simulated individuals whose value for the unobserved characteristic is greater than 2.

5.2.3 Revealing Demand

If we are able to observe a large number of prices, then we can use revealed preference to estimate the whole distribution of the unobserved utility. At each price, the share of individuals purchasing product A is calculated. If we observe enough prices then we can use the observed shares at each price to plot out the demand curve.

```
> p <- runif(9,min=-10,max=10)
> # 9 points between -10, 10.
> s <- matrix(NA,length(p),1)  # share of market buying A.
> for (i in 1:length(p)) {
+    s[i,1] <- mean(u - p[i] > 0)
+    #print(i)
+ }
```

Figure 5.1 presents the estimated demand curve.

5.3 Discrete Choice

Demand estimation often involves outcomes with discrete values. In McFadden's original problem, we observe one of three choices, car, bus, or train. OLS tends not to work very well when the outcome of interest is discrete or limited in some manner. Given this, it may be preferable to use a discrete choice model such as a logit or probit.[4]

The section uses simulated data to illustrate issues with estimating the discrete choice model.

---

[4]In machine learning, discrete choice is referred to as a classification problem.

```
> plot(1-ecdf(u)(u),u, type="l",lwd=3,lty=1,col=1,
+       xlab="s", ylab="p", xlim=c(0,1))
> # ecdf(a)(a) presents the estimated probabilities of a.
> lines(sort(s),p[order(s)], type="l", lwd=3,lty=2)
> abline(h=0, lty=2)
> legend("bottomleft",c("True","Est."),lwd=3,lty=c(1:2))
```

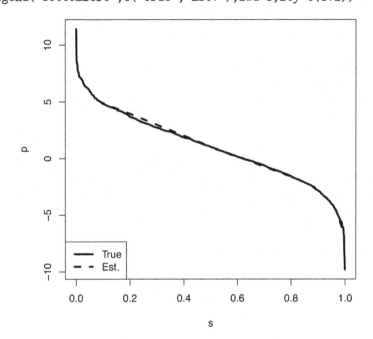

FIGURE 5.1
Plot of survival function and estimated demand.

5.3.1   Simple Discrete Choice Model

Consider the following discrete model. There is some **latent** (hidden) value of
the outcome (y_i^*), where if this value is large enough we observe $y_i = 1$.

$$y_i^* = a + bx_i + v_i$$

$$y_i = \begin{cases} 1 \text{ if } y_i^* \geq 0 \\ 0 \text{ if } y_i^* < 0 \end{cases} \tag{5.2}$$

We can think of y_i^* as representing **utility** and y_i as representing observed
demand.

5.3.2   Simulating Discrete Choice

In the simulated data there is a latent variable (y^*) that is determined by a
model similar to the one presented in Chapter 1.[5] Here, however, we do not
observe y^*. We observe y which only has values 0 or 1.

```
> set.seed(123456789)
> N <- 100
> a <- 2
> b <- -3
> u <- rnorm(N)
> x <- runif(N)
> y_star <- a + b*x + u
> y <- y_star > 0
> lm1 <- lm(y ~ x)
```

Figure 5.2 shows that the estimated relationship differs substantially from
the true distribution. The figure illustrates how OLS fails to accurately estimate
the parameters of the model. In order to correctly estimate the relationship
we need to know the distribution of the unobserved term.

5.3.3   Modeling Discrete Choice

We can write out the model using matrix notation.

$$y_i^* = \mathbf{X}_i'\beta + v_i$$

$$y_i = \left\{ \begin{array}{l} 1 \text{ if } y_i^* \geq 0 \\ 0 \text{ if } y_i^* < 0 \end{array} \right. \tag{5.3}$$

where $\mathbf{X}_i$ is a vector of observed characteristics of individual i, β is a vector
that maps from the observed characteristics to the **latent outcome**, y_i^*, and v_i
is the unobserved term. The observed outcome, y_i is equal to 1 if the **latent
variable** is greater than 0 and it is 0 if the **latent variable** is less than 0.

The probability of observing one of the outcomes $(y_i = 1)$ is as follows.

$$\begin{aligned} \Pr(y_i = 1 | \mathbf{X}_i) \quad &= \Pr(y_i^* > 0) \\ &= \Pr(\mathbf{X}_i'\beta + v_i > 0) \\ &= \Pr(v_i > -\mathbf{X}_i'\beta) \\ &= 1 - F(-\mathbf{X}_i'\beta) \end{aligned} \tag{5.4}$$

where F represents the probability distribution function of the unobserved
characteristic.

If we know β, we can determine the distribution of the unobserved term
with variation in the Xs. That is, we can determine F. This is exactly what

---

[5]This is actually a probit model presented in more detail below.

```
> plot(x,y, ylim=c(-0.2,1))
> abline(a = 2,b = -3, lwd=2)
> abline(a = lm1$coefficients[1], b=lm1$coefficients[2],
+          lty=2, lwd=2)
> legend("bottomleft", c("True", "Est."), lwd=2, lty=1:2)
```

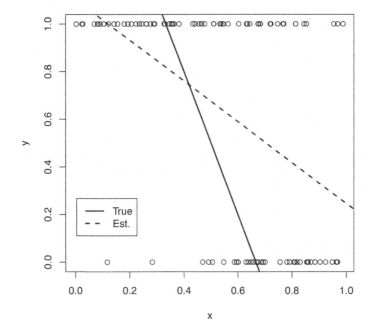

FIGURE 5.2
Plot of x and y with the relationship between y^* and x represented by the solid line. The estimated relationship is represented by the dashed line.

we did in the previous section. In that section, $\beta = 1$ and variation in the price (the Xs) determines the probability distribution of u. However, we do not know β. In fact, β is generally the policy parameter we want to estimate. We cannot identify both β and F without making some restriction on one or the other.

The standard solution is to assume we know the true distribution of the unobserved term (F). In particular, a standard assumption is to assume that F is **standard normal**. Thus, we can write the probability of observing the outcome $y_i = 1$ conditional on the observed Xs.

$$\Pr(y_i = 1 | \mathbf{X}_i) = 1 - \Phi(-\mathbf{X}_i'\beta) \tag{5.5}$$

where Φ is standard notation for the standard normal distribution function.

5.4   Maximum Likelihood

The standard algorithm for estimating discrete choice models is **maximum likelihood**. The maximum likelihood algorithm generally requires some assumption about the distribution of the error term. However, as seen above, we are making such an assumption anyway.

This section takes a detour to illustrate how the maximum likelihood algorithm works.

5.4.1   Binomial Likelihood

Consider the problem of determining whether a coin is "fair." That is, whether the coin has an equal probability of Heads or Tails, when tossed. If the coin is weighted then it may not be fair. It may have a greater probability of landing on Tails than on Heads. The code simulates an unfair coin. The observed probability of a head is 34 of 100.

```
> set.seed(123456789)
> N <- 100
> p <- 0.367 # the true probability of Head.
> Head <- runif(N) < p
> mean(Head) # the observed frequency of Head.
```

```
[1] 0.34
```

What is the likelihood that this data was generated by a fair coin? It is the probability of observing 34 Heads and 66 Tails given that the true probability of a Head is 0.5.

What is the probability of observing 1 Head given the true probability is 0.5? It is just the probability of Heads, which is 0.5.

$$\Pr(\text{Head}|p = 0.5) = 0.5 \tag{5.6}$$

What is the probability of observing three Heads and zero Tails? If the coin tosses are independent of each other, then it is the probability of each Head, all multiplied together.[6]

$$\Pr(\{\text{Head}, \text{Head}, \text{Head}\}|p = 0.5) = 0.5 \times 0.5 \times 0.5 = 0.5^3 \tag{5.7}$$

How about three Heads and two Tails?

$$\Pr(\{3\text{Heads}, 2\text{Tails}\}) = 0.5^3 0.5^2 \tag{5.8}$$

---

[6]Independent means that if I know the first two coin tosses result in Heads, the probability of a Head in the third coin toss is the same as if I saw two Tails or any other combination. The previous coin toss provides no additional information about the results of the next coin toss, if the true probability is known.

Actually, it isn't. This is the probability of observing 3 Heads and *then* 2 Tails. But it could have been 1 Head, 2 Tails, 2 Heads or 1 Tail, 1 Head, 1 Tail, 2 Heads etc., etc. There are a number of different combinations of results that have 3 Heads and 2 Tails. In this case there are $\frac{5!}{3!2!} = 10$ different permutations, where 5! means 5 factorial or 5 multiplied by 4 multiplied by 3 multiplied by 2 multiplied by 1.

In **R** we can use `factorial()` to do the calculation.

```
> factorial(5)/(factorial(3)*factorial(2))
```

```
[1] 10
```

What is the likelihood of observing 34 Heads and 66 Tails? If the true probability is 0.5, the likelihood is given by the binomial function.

$$\Pr(\{34H, 66T\}|p = 0.5) = \frac{100!}{34!66!}0.5^{34}0.5^{66} \tag{5.9}$$

What is the likelihood of observing $\hat{p}$ Heads in N trials? Given a true likelihood of p, it is given by the binomial function.

$$\Pr(\hat{p}|p, N) = \frac{N!}{(\hat{p}N)!((1 - \hat{p})N)!}p^{\hat{p}N}(1 - p)^{(1-\hat{p})N} \tag{5.10}$$

In **R** we can use the `choose()` function to calculate the coefficient for the binomial function.

```
> choose(100, 34)*(0.5^100)
```

```
[1] 0.0004581053
```

The likelihood that the coin is fair seems small.

What is the most likely true probability? One method uses the **analogy principle**. If we want to know the true probability then we use the analogy in the sample. The best estimate of the true probability is the observed frequency of Heads in the sample (Manski, 1990).[7] It is $34/100$. Note that this is not equal to the true probability of 0.367, but it is pretty close.

Alternatively, find the probability that maximizes the likelihood. What is the true probability p that has the highest likelihood of generating the observed data? It is the true probability that maximizes the following problem.

$$\max_{p \in [0,1]} \quad \frac{N!}{(\hat{p}N)!((1-\hat{p})N)!}p^{\hat{p}N}(1 - p)^{(1-\hat{p})N} \tag{5.11}$$

It is not a great idea to ask a computer to solve the problem as written. The issue is that these numbers can be very very small. Computers have a tendency to change very small numbers into other, totally different, small numbers. This can lead to errors.

---

[7]See discussion of the analogy principle in Appendix A.

Find the probability that maximizes the log likelihood.

$$\max_{p\in[0,1]} \quad \begin{aligned} &\log(N!) - log((\hat{p}N)!) - \log((1-\hat{p})N)!) \\ &+\hat{p}N\log(p) + (1-\hat{p})N\log(1-p) \end{aligned} \tag{5.12}$$

The solution to this problem is identical to the solution to the original problem.[8]

5.4.2 Binomial Likelihood in R

Equation (5.12) provides pseudo-code for a simple optimizer in **R**. We can use the `optimize()` function to find the minimum value in an interval.[9] Note also that the function being optimized dropped the coefficient of the binomial function. Again, this is fine because the optimum does not change.

```
> function_binom <- function(p, N, p_hat) {
+    return(-((p_hat*N)*log(p) + (1 - p_hat)*N*log(1-p)))
+    # Note the negative sign as optimize is a minimizer.
+ }
> optimize(f = function_binom, interval=c(0,1), N = 100,
+          p_hat=0.34)

$minimum
[1] 0.3399919

$objective
[1] 64.10355
```

The maximum likelihood estimate is 0.339, which is fairly close to the analog estimate of 0.34. Figure 5.3 shows that the likelihood function is relatively flat around the true value. The implication is that the difference in the likelihood between the true value and the estimated value is quite small.

```
> p <- c(1:1000)/1000
> log_lik <- -function_binom(p, N=100, p_hat=0.34)
> # note that the function acts on the whole vector.
> # negative to show maximum likelihood.
```

5.4.3 OLS with Maximum Likelihood

We can use maximum likelihood to estimate OLS. Chapter 1 presented the standard algorithms for estimating OLS. It points out that with an additional assumption, the maximum likelihood algorithm could also be used instead.

---

[8]Optimums do not vary with monotonic transformations.

[9]The function `optimize()` is used when optimizing over one variable, while `optim()` is used for optimizing over multiple variables.

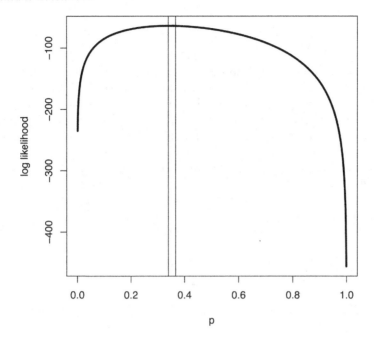

FIGURE 5.3
Plot of log likelihood for a sample with 34 Heads out of 100. Lines at the analog estimate of 0.34 and the true value of 0.367.

Assume we have data generated by the following linear model.

$$y_i = \mathbf{X}_i'\beta + v_i \tag{5.13}$$

where y_i is the outcome of interest, $\mathbf{X}_i$ is a vector representing observed characteristics of individual i, and v_i represents the unobserved characteristics of individual i. As is normally the case, we are interested in estimating β.

We can use maximum likelihood to estimate parameters of this model. However, we must assume that v_i has a particular distribution. A standard assumption is that $v_i \sim \mathcal{N}(0, \sigma^2)$. That is, we assume the unobserved characteristic is normally distributed. Note that we don't know the parameter, σ.

We can determine the likelihood of observing the data by first rearranging Equation (5.13).

$$v_i = y_i - \mathbf{X}_i'\beta \tag{5.14}$$

The probability of observing the outcome is as follows.

$$
\begin{aligned}
f(v_i|y_i, \mathbf{X}_i) &= \tfrac{1}{\sigma}\phi\left(\tfrac{v_i - 0}{\sigma}\right) \\
&= \tfrac{1}{\sigma}\phi(z_i)
\end{aligned}
\tag{5.15}
$$

where

$$z_i = \frac{y_i - \mathbf{X}'_i\beta}{\sigma} \tag{5.16}$$

and ϕ is the **standard normal density**.

Note in **R**, it is necessary to use the **standard normal distribution** function.[10] To use this function we need to **normalize** the random variable by taking away the mean of the unobserved term, which is zero, and by dividing by the standard deviation of the unobserved term (σ). The standard notation for the normalized variable is z, but don't confuse this with our notation for an instrumental variable. We also need to remember that this is the density, a derivative of the probability distribution function. Therefore, we need to adjust the density formula by dividing it by the standard deviation of the distribution of the unobserved characteristic (σ).

Therefore, the likelihood of observing the data is given by the following product.

$$L(\{y, \mathbf{X}\}|\{\beta, \sigma\}) = \prod_{i=1}^{N} \left(\frac{1}{\sigma}\phi\left(\frac{y_i - \mathbf{X}'_i\beta}{\sigma}\right)\right) \tag{5.17}$$

The sample size is N and $\prod_{i=1}^{N}$ is notation for multiply all the items denoted 1 to N together.[11]

We can find the maximum likelihood estimates of β and σ by solving the following problem.

$$\max_{\hat{\beta},\hat{\sigma}} \sum_{i=1}^{N} \log\left(\phi\left(\frac{y_i - \mathbf{X}'_i\hat{\beta}}{\hat{\sigma}}\right)\right) - N\log(\hat{\sigma}) \tag{5.18}$$

Compare this to the estimator in Chapter 1.[12]

5.4.4 Maximum Likelihood OLS in R

We can create a maximum likelihood estimator of the OLS model using Equation (5.18) as pseudo-code.

```
> f_ols_ml <- function(par, y, X) {
+    X <- cbind(1,X)
+    N <- length(y)
+    J <- dim(X)[2]
+    sigma <- exp(par[1])
+    # Note that sigma must be positive.
+    # The exponential function maps
+    # from any real number to positive numbers.
```

---

[10]This has to do with the ability of this function to quickly run through vectors. See Appendix B for a discussion of programming in **R**.

[11]This assumes that the outcomes are independent and identically distributed.

[12]See the formula for standard normal.

```
+    # It allows the optimizer to choose any value and
+    # transforms that number into a positive value.
+    beta <- par[2:(J+1)]
+    z <- (y - X%*%beta)/sigma
+    log_lik <- -sum(log(dnorm(z)) - log(sigma))
+    return(log_lik)
+    # remember we are minimizing.
+ }
```

The standard optimizer in **R** is the function `optim()`. This function defaults to a Nelder-Mead which is a fairly robust algorithm.

```
> a <- optim(par=c(0,2,-3),fn=f_ols_ml,y=y_star,X=x)
> # optim takes in starting values with par, then the function
> # used and then values that the function needs.
> # we cheat by having it start at the true values.
> # sigma
> exp(a$par[1])
```

```
[1] 0.9519634
```

```
> # beta
> a$par[2:3]
```

```
[1]   1.832333 -2.643751
```

Our estimate $\hat{\sigma}$ fairly close to the true value is 1, and $\hat{\beta} = \{1.83, -2.64\}$ compared to true values of 2 and -3. What happens if you use different starting values?

5.4.5 Probit

Circling back, consider the discrete choice problem that began the chapter. If we have information about the distribution of the unobserved term, which is generally assumed, then we can find the parameters that maximize the likelihood of the model predicting the data we observe.

Consider the problem described by Equation (5.3). Assume the unobserved characteristic is distributed standard normal, then the likelihood of observing the data is given by the following function.

$$L(\{y, \mathbf{X}\}|\beta) - \prod_{i=1}^{N} \Phi(-\mathbf{X}_i'\beta)^{1-y_i}(1 - \Phi(-\mathbf{X}_i'\beta))^{y_i} \tag{5.19}$$

Equation (5.19) shows the likelihood function of observing the data we actually observe ($\{y, \mathbf{X}\}$) given that the true probability which is determined by β. The variance of the unobserved characteristic (σ^2) is not identified in the data.

That is, there are an infinite number of values for σ^2 that are consistent with the observed data. The standard solution is to set it equal to 1 ($\sigma^2 = 1$). Note that y_i is either 0 or 1 and also note that $a^0 = 1$, while $a^1 = a$. This is the binomial likelihood function (Equation (5.11)) written in a general way with the probabilities determined by the standard normal function.

The parameter of interest, β, can be found as the solution to the following maximization problem.

$$\max_{\hat{\beta}} \sum_{i=1}^{N} (1 - y_i) \log(\Phi(-\mathbf{X}_i'\hat{\beta})) + y_i \log(\Phi(\mathbf{X}_i'\hat{\beta}))) \qquad (5.20)$$

I made a slight change going from Equation (5.19) to Equation (5.20). I took advantage of the fact that the **normal distribution** is symmetric. This version is better for computational reasons.[13]

5.4.6 Probit in R

```
> f_probit <- function(beta, y, X) {
+     X <- cbind(1,X)
+     Xb <- X%*%beta
+     log_lik <- (1 - y)*log(pnorm(-Xb)) + y*log(pnorm(Xb))
+     return(-sum(log_lik))
+ }
> optim(par=lm1$coefficients,fn=f_probit,y=y,X=x)$par

(Intercept)            x
   2.014153    -2.835234
```

We can use Equation (5.20) as the basis for our own probit estimator. The probit estimates are closer to the true values of 2 and -3, although they are not particularly close to the true values. Why aren't the estimates close to the true values?[14]

5.4.7 Generalized Linear Model

The probit is an example of a **generalized linear model**. The outcome vector is, $y = f(\mathbf{X}\beta)$, where f is some function. This is a generalization of the OLS model. It has a **linear index**, $\mathbf{X}\beta$, but that index sits inside a potentially non-linear function (f). The probit is an example, as is the logit model discussed below.

In **R** these types of functions can often be estimated with the glm() function. Like the lm() function, glm() creates an object that includes numerous results including the coefficients. The nice thing about the glm() function is that it

---

[13] Thanks to Joris Pinkse for pointing this issue out.
[14] This issue is discussed in Chapter 1.

includes a variety of different models. Unfortunately, that makes it unwieldy to use.

We can compare our probit estimates to those from the built-in **R** probit model using `glm()`.

```
> glm(y ~ x, family = binomial(link="probit"))$coefficients

(Intercept)             x
  2.014345      -2.835369
```

The results are about the same. The two models are solved using different algorithms. The `glm()` uses an algorithm called iterative weighted least squares rather than maximum likelihood.

5.5 McFadden's Random Utility Model

In order to estimate the impact of the BART rail system, McFadden needed a model that captured current choices and predicted demand for a product that didn't exist.

The section presents McFadden's model, the probit, and logit estimators and simulation results.

5.5.1 Model of Demand

In McFadden's model, person i's utility over choice j is the following random function.

$$U_{ij} = \mathbf{X}'_{ij}\beta + v_{ij} \tag{5.21}$$

Person i's utility is a function of observable characteristics of both person i and the choice j represented in Equation (5.21) by the matrix $\mathbf{X}$. In the case of the BART survey, these are things like the person's income and the cost of commuting by car. These observed characteristics are mapped into utility by person i's preferences, represented by the β vector. Lastly, there are unobserved characteristics of person i and choice j that also determine the person's value for the choice. These are represented by v_{ij}.

To predict demand for BART from observed demand for cars and buses we need two assumptions. First, the preference weights (β) cannot vary with the product. Second, the choice can be described as a basket of characteristics. This style of model is often called an **hedonic model**. Commuting by car is represented by a set of characteristics like tolls to be paid, gasoline prices, parking, the time it takes, and car maintenance costs. Similarly, commuting by train or bus depends on the ticket prices and the time it takes. Individuals weight these characteristics the same irrespective of which choice they refer to.

We can use **revealed preference** and observed choices to make inferences about person i's preferences. From the revealed preference assumption we learn that the person's utility from product A is greater than their utility from product B. If we observe person i face the choice between two products A and B, and we see her choose A, then we learn that $U_{iA} > U_{iB}$.

$$U_{iA} > U_{iB}$$
$$\mathbf{X}'_{iA}\beta + v_{iA} > \mathbf{X}'_{iB}\beta + v_{iB} \qquad (5.22)$$
$$v_{iA} - v_{iB} > -(\mathbf{X}_{iA} - \mathbf{X}_{iB})'\beta$$

Equation (5.22) shows that if there is enough variation in the observed characteristics of the choices $(\mathbf{X}_{iA} - \mathbf{X}_{iB})$, we can potentially estimate the unobserved characteristics $(v_{iA} - v_{iB})$ and the preferences (β). The analysis in the first section of the chapter shows how variation in prices can allow the distribution of the unobserved characteristic to be mapped out. The Xs are like the price; in fact, they may be prices. We also need to make an assumption about the distribution of the unobserved characteristics $(v_{iA} - v_{iB})$. Next we consider two different assumptions.

5.5.2   Probit and Logit Estimators

If we assume that $v_{iA} - v_{iB}$ is distributed **standard normal** we can use a probit model.[15]

$$
\begin{aligned}
\Pr(y_i = 1 | \mathbf{X}_{iA}, \mathbf{X}_{iB}) &= \Pr(v_{iA} - v_{iB} > -(\mathbf{X}_{iA} - \mathbf{X}_{iB})'\beta) \\
&= \Pr(-(v_{iA} - v_{iB}) < (\mathbf{X}_{iA} - \mathbf{X}_{iB})'\beta) \qquad (5.23) \\
&= \Phi((\mathbf{X}_{iA} - \mathbf{X}_{iB})'\beta)
\end{aligned}
$$

McFadden's original paper estimates a logit. It assumes that the unobserved characteristics are distributed extreme value type 1. This mouthful-of-a-distribution is also called Gumbel or log Weibull. The advantage of this distribution is that the difference in the unobserved terms is a logistic distribution and a logit model can be used.

The logit has some very nice properties. In particular, it is very easy to compute. This made it a valuable model in the 1970s. Even today the logit is commonly used in machine learning because of its computational properties.[16] The logit assumption allows the probability of interest to have the following form.

$$\Pr(y_i = 1 | \mathbf{X}_{iA}, \mathbf{X}_{iB}) = \frac{\exp((\mathbf{X}_{iA} - \mathbf{X}_{iB})'\beta)}{1 + \exp((\mathbf{X}_{iA} - \mathbf{X}_{iB})'\beta)} \qquad (5.24)$$

This function is very useful. It has the property that whatever parameter

---

[15]Note in the simulated data, the v_{iA} term is dropped so that the assumption holds.

[16]It is used in neural network estimation models. In machine learning this is called a "sigmoid" function.

you give it, it returns a number between 0 and 1, a probability.[17] It is often used in optimization problems for this reason.

5.5.3 Simulation with Probit and Logit Estimators

Consider a simulation of the McFadden model with both the probit and logit assumption on the unobserved characteristics.

```
> set.seed(123456789)
> N <- 5000
> X_A <- cbind(1,matrix(runif(2*N),nrow=N))
> X_B <- cbind(1,matrix(runif(2*N),nrow=N))
> # creates two product characteristic matrices
> beta <- c(1,-2,3)
```

In the simulation there are 5,000 individuals choosing between two products with two observable characteristics. Note that these characteristics vary across the individuals, but the preferences of the individuals do not.

```
> # Probit
> u_A <- rnorm(N)
> y <- X_A%*%beta - X_B%*%beta + u_A > 0
> glm1 <- glm(y ~ I(X_A - X_B),
+                 family = binomial(link="probit"))
> # note that I() does math inside the glm() function.
> # g "el" m one.
```

The probit model assumes that the unobserved characteristic (the relative unobserved characteristic) is distributed **standard normal**. The logit assumes that the unobserved characteristics are distributed extreme value type 1. Note that the function I() allows mathematical operations within the glm() or lm() function. Here it simply takes the difference between the two matrices of observed characteristics.

```
> # Logit
> u_A <- log(rweibull(N, shape = 1))
> # gumbel or extreme value type 1
> u_B <- log(rweibull(N, shape = 1))
> y <- (X_A%*%beta - X_B%*%beta) + (u_A - u_B) > 0
> glm2 <- glm(y ~ I(X_A - X_B), family = binomial(link="logit"))
```

What happens if you run OLS? Do you get the right sign? What about magnitude?

Table 5.1 presents the probit and logit estimates. The table shows that

---

[17]It is this property that makes it useful as an "activation" function in neural network models.

	Dependent variable:	
	y	
	probit	*logistic*
	(1)	(2)
I(X_A - X_B)1		
I(X_A - X_B)2	−2.023***	−1.924***
	(0.068)	(0.089)
I(X_A - X_B)3	3.016***	2.861***
	(0.081)	(0.099)
Constant	0.010	−0.018
	(0.023)	(0.033)
Observations	5,000	5,000
Note:	*p<0.1; **p<0.05; ***p<0.01	

TABLE 5.1
Probit estimate (1) and logit estimate (2).

the probit gives estimates that are very close to the true values of -2 and 3. Why is the intercept term 0 rather than the true value of 1? The estimates from the logit are also relatively close to the true values. The different error assumptions of the logit may lead to wider variation in the estimates.

5.6 Multinomial Choice

In the analysis above, individuals choose between two options. In many problems individuals have many choices. This section looks at commuters choosing between car, bus, and train. It is in these multinomial choice problems that the logit really comes into its own.

The section presents a multinomial probit, actually a bivariate probit which can be used to model the choice between three options. This model is relatively general, but the computational burden increases exponentially (possibly geometrically) in the number of choices. This is called the **curse of dimensionality**. One solution to this computational issue is to use an ordered probit. The book doesn't discuss this model but the model can be very useful for certain problems.

Instead, the section considers a multinomial logit and the **independence of irrelevant alternatives** assumption. The assumption implies that the unobserved characteristics associated with one choice are independent of the unobserved characteristics associated with any other choice. This assumption allows the logit model to handle very large choice problems. It also allows the model to handle predictions about new choices that are not in the observed data, such as the BART rail system. However, it is a strong restriction on preferences.

The section illustrates these methods with simulated data.

5.6.1 Multinomial Choice Model

Consider the full model of choice over three potential modes of transportation. In the model each person i has utility for each of the three choices. Note that as before, β is the same for each choice. It is the observed characteristics of the choices and the unobserved characteristics that change between choices.

$$
\begin{aligned}
U_{iC} &= \mathbf{X}'_{iC}\beta + v_{iC} \\
U_{iT} &= \mathbf{X}'_{iT}\beta + v_{iT} \\
U_{iB} &= \mathbf{X}'_{iB}\beta + v_{iB}
\end{aligned}
\tag{5.25}
$$

where U_{iC} is the utility that individual i receives from choosing car, U_{iT} is the utility for train and U_{iB} is the utility for bus. In the case where there is no rail, person i chooses a car if their utility from a car is higher than utility from a bus.

$$U_{iC} > U_{iB}$$
$$-(v_{iC} - v_{iB}) < (\mathbf{X}_{iC} - \mathbf{X}_{iB})'\beta \tag{5.26}$$

$$y_{iC} = \begin{cases} 1 \text{ if } (\mathbf{X}_{iC} - \mathbf{X}_{iB})'\beta + (v_{iC} - v_{iB}) > 0 \\ 0 \text{ if } (\mathbf{X}_{iC} - \mathbf{X}_{iB})'\beta + (v_{iC} - v_{iB}) < 0 \end{cases} \tag{5.27}$$

In the case where rail is also a choice, person i will choose a car if and only if their utility for car is greater than both bus and train.

$$(\mathbf{X}_{iC} - \mathbf{X}_{iB})'\beta + (v_{iC} - v_{iB}) > \max\{0, (\mathbf{X}_{iT} - \mathbf{X}_{iB})'\beta + (v_{iT} - v_{iB})\} \tag{5.28}$$

As discussed above, the standard approach to estimation of choice problems in economics is to have a "left out" choice or reference category. In this case, that transport mode is bus. The observed characteristics of car and train are created in reference to bus. Let $\mathbf{W}_{iC} = \mathbf{X}_{iC} - \mathbf{X}_{iB}$ and $\mathbf{W}_{iT} = \mathbf{X}_{iT} - \mathbf{X}_{iB}$.

From Equation (5.28), the probability of observing the choice of car is as a follows.

$$\Pr(y_{iC} = 1 | \mathbf{W}_{iC}, \mathbf{W}_{iT}) = \begin{cases} (v_{iC} - v_{iB}) > -\mathbf{W}'_{iC}\beta \text{ and} \\ (v_{iC} - v_{iT}) > -(\mathbf{W}_{iC} - \mathbf{W}_{iT})'\beta \end{cases} \tag{5.29}$$

Figure 5.4 depicts the three choices as a function of the unobserved characteristics. If the relative unobserved characteristics of *both* choices are low enough, then neither is chosen. That is, the individual chooses Bus. The individual chooses Car if the relative unobserved characteristic is higher for Car than it is for Train.

5.6.2 Multinomial Probit

When there are three choices, the multinomial normal is a bivariate normal. The bivariate probit model assumes $\{v_{iC} - v_{iB}, v_{iT} - v_{iB}\} \sim \mathcal{N}(\mu, \boldsymbol{\Sigma})$, where $\mu = \{0, 0\}$ and

$$\boldsymbol{\Sigma} = \begin{bmatrix} 1 & \rho \\ \rho & 1 \end{bmatrix} \tag{5.30}$$

and $\rho \in [-1, 1]$ is the correlation between the unobserved characteristics determining the relative demand for cars and trains. Note that the variance of the relative unobserved characteristics is not identified by the choice data and is assumed to be 1.

The simplest case is the probability of observing Bus ($\{0, 0\}$).

$$\Pr(y_i = \{0, 0\} | \mathbf{W}, \beta, \rho) = \int_{-\infty}^{-\mathbf{W}'_{iC}\beta} \Phi\left(\frac{-\mathbf{W}'_{iT}\beta - \rho v_1}{\sqrt{(1 - \rho^2)}}\right) \phi(v_1)dv_1 \tag{5.31}$$

Simplest, not simple. A word to the wise, writing out multivariate probits is a good way to go insane.

There is a lot to unpack in Equation (5.31). First, it is written out in a way

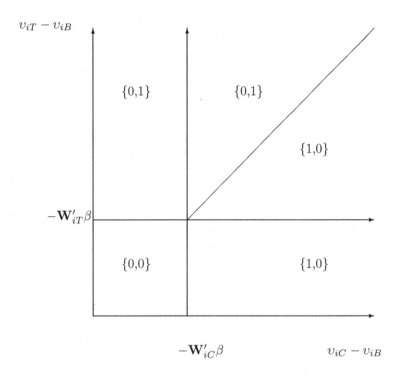

FIGURE 5.4
Choice over car $\{1,0\}$, bus $\{0,0\}$ and train $\{0,1\}$.

that is going to allow "easy" transition to **R** code. The probability in question is a joint probability over the two relative unobserved characteristics. This joint probability can be written as a mixture distribution of the probability of not choosing Train conditional on the unobserved characteristic for the choice of Car. Next, the probability over not choosing Train is written out using a standard normal distribution. This is useful for programming it up in **R**. Lastly, because the unobserved characteristics may be correlated, the variable needs to be **normalized**, where the distribution of the conditional unobserved characteristic has a mean of ρv_1 and a variance of $1 - \rho^2$.[18]

The probability of observing the choice of Car ($\{1,0\}$) is similar, although even more complicated. Consider Figure 5.4. We are interested in measuring the trapezoid that is below the diagonal line and to the right of the vertical

---

[18]See Goldberger (1991) for the best exposition that I am aware of.

line.

$$\Pr(y_i = \{1, 0\} | \mathbf{W}, \beta, \rho) = \int_{-\mathbf{W}'_{iC}\beta}^{\infty} \Phi\left(\frac{(\mathbf{W}_{iT} - \mathbf{W}_{iC})'\beta + (1 - \rho)v_1}{\sqrt{(1 - \rho^2)}}\right) \phi(v_1)dv_1$$

(5.32)

Now the choice of Car is determined by both the relative value of Car to Bus and the relative value of Car to Train. If the unobserved characteristic associated with the Car choice is high, then the unobserved characteristic associated with Train does not have to be that low for Car to be chosen.

5.6.3 Multinomial Probit in R

Equations (5.31) and (5.32) have been written in a way to suggest pseudo-code for the bivariate probit. Note how the integral is done. The program takes a set of draws from a standard normal distribution and then for each draw it calculates the appropriate probabilities. After the loop has finished, the probabilities are averaged. This approach gives relatively accurate estimates even with relatively small numbers of loops. Even so, the computational burden of this estimator is large. Moreover, this burden increases dramatically with each choice that is added to the problem.

```
> f_biprobit <- function(par, y, W1, W2, K = 10) {
+    # K is a counter for the number of draws from the normal
+    # distribution - more draws gives greater accuracy
+    # but slower computational times.
+    # function setup
+    set.seed(123456789)  # helps the optimizer work better.
+    epsilon <- 1e-20  # used below to make sure the logs work.
+    W1 <- cbind(1,W1)
+    W2 <- cbind(1,W2)
+    N <- dim(y)[1] # note y is a matrix.
+    J <- dim(W1)[2]
+    rho <- exp(par[1])/(1 + exp(par[1]))
+    # the "sigmoid" function that keeps value between 0 and 1.
+    # It assumes that the correlation is positive.
+    beta <- par[2:(J+1)]
+    # integration to find the probabilities
+    u <- rnorm(K)
+    p_00 <- p_10 <- rep(0, N)
+    for (u_k in u) {
+       u_k0 <- u_k < -W1%*%beta
+       p_00 <- p_00 +
+          u_k0*pnorm((-W2%*%beta - rho*u_k)/((1 - rho^2)^(.5)))
+       p_10 <- p_10 + (1 - u_k0)*pnorm(((W1 - W2)%*%beta +
+                                     (1 - rho)*u_k)/((1 - rho^2)^(.5)))
```

```
+   }
+   # determine the likelihood
+   log_lik <- (y[,1]==0 & y[,2]==0)*log(p_00/K + epsilon) +
+     (y[,1]==1 & y[,2]==0)*log(p_10/K + epsilon) +
+     (y[,1]==0 & y[,2]==1)*log(1 - p_00/K - p_10/K + epsilon)
+   return(-sum(log_lik))
+ }
```

5.6.4 Multinomial Logit

Given the computational burden of the multivariate normal, particularly as the number of choices increase, it is more common to assume a multinomial logit. This model assumes that the unobserved characteristic affecting the choice between Car and Bus is independent of the unobserved characteristic affecting the choice between Car and Train.[19]

$$\Pr(y_i = \{1,0\}|\mathbf{W}, \beta) = \frac{\exp(\mathbf{W}'_{iC}\beta)}{1 + \exp(\mathbf{W}'_{iC}\beta) + exp(\mathbf{W}'_{iT}\beta)} \tag{5.33}$$

As above, a maximum log likelihood algorithm is used to estimate the multinomial logit.

5.6.5 Multinomial Logit in R

Starting with the bivariate probit and Equation (5.33), we can write out the multinomial logit with three choices.

```
> f_logit <- function(beta,y,W1,W2) {
+   epsilon <- 1e-20 # so we don't take log of zero.
+   W1 <- as.matrix(cbind(1,W1))
+   W2 <- as.matrix(cbind(1,W2))
+   W1b <- W1%*%beta
+   W2b <- W2%*%beta
+   p_10 <- exp(W1b)/(1 + exp(W1b) + exp(W2b))
+   p_01 <- exp(W2b)/(1 + exp(W1b) + exp(W2b))
+   p_00 <- 1 - p_10 - p_01
+   log_lik <- (y[,1]==0 & y[,2]==0)*log(p_00 + epsilon) +
+     (y[,1]==1 & y[,2]==0)*log(p_10 + epsilon) +
+     (y[,1]==0 & y[,2]==1)*log(p_01 + epsilon)
+   return(-sum(log_lik))
+ }
```

---

[19]The logit function has the advantage that it is very easy to estimate, but the disadvantage that it places a lot of restrictions on how preferences determine choices. Given these restrictions there is interest in using more flexible models such as the multi-layered logit (deep neural-network) or the mixed logit model discussed in Nevo (2000).

Notice that, unlike the multinomial probit, we do not need to take random number draws. Calculating the probabilities is much simpler and computationally less intensive.

5.6.6 Simulating Multinomial Choice

Now we can compare the two methods. The simulation uses a bivariate normal distribution to simulate individuals choosing between three options. The simulation also assumes that the unobserved characteristics are correlated between choices.

```
> require(mvtnorm)
> # this package creates multivariate normal distributions
> set.seed(123456789)
> N <- 1000
> mu <- c(0,0)
> rho <- 0.1 # correlation parameter
> Sigma <- cbind(c(1,rho),c(rho,1))
> u <- rmvnorm(N, mean = mu, sigma = Sigma)
> # relative unobserved characteristics of two choices
> x1 <- matrix(runif(N*2), nrow=N)
> x2 <- matrix(runif(N*2), nrow=N)
> # creates matrix of N x 2 random numbers.
> # relative observed characteristics of two choices.
> a <- -1
> b <- -3
> c <- 4
> U <- a + b*x1 + c*x2 + u
> y <- matrix(0, N, 2)
> y[,1] <- U[,1] > 0 & U[,1] > U[,2]
> y[,2] <- U[,2] > 0 & U[,2] > U[,1]
> par <- c(log(rho),a,b,c)
> W1 <- cbind(x1[,1], x2[,1])
> W2 <- cbind(x1[,2], x2[,2])
> start_time <- Sys.time()
> a1 <- optim(par=par, fn=f_biprobit,y=y,W1=W1,W2=W2,
+             K=100,control = list(trace=0,maxit=1000))
> # time of f_biprobit
> Sys.time() - start_time

Time difference of 7.292156 secs

> par <- c(a,b,c)
> start_time <- Sys.time()
> b1 <- optim(par=par, fn=f_logit,y=y,W1=W1,W2=W2,
+             control = list(trace=0,maxit=1000))
```

```
> # time of f_logit
> Sys.time() - start_time
```

Time difference of 0.05458403 secs

```
> a1$par[2:4]
```

[1] -1.039715 -2.819453 3.887579

```
> b1$par[1:3]
```

[1] -1.591007 -4.976470 6.687181

At 100 draws, the bivariate probit estimator does pretty well, with $\hat{\beta} = \{-1.04, -2.82, 3.89\}$ compared to $\beta = \{-1, -3, 4\}$. The logit model doesn't do nearly as well. This isn't surprising given that the assumptions of the model do not hold in the simulated data. Even still it gets the signs of the parameters and you can see that it solves much quicker than the bivariate probit.

5.7 Demand for Rail

In McFadden's analysis, policy makers were interested in how many people would use the new BART rail system. Would there be enough users to make such a large infrastructure project worthwhile? This question remains relevant for cities across the US. Many large US cities like New York, Chicago and Boston, have major rail infrastructure, while many smaller US cities do not. For these smaller cities, the question is whether building a substantial rail system will lead to an increase in public transportation use.

To analyze this question we can use data from the National Household Travel Survey.[20] In particular, the publicly available household component of that survey. The data provides information on what mode of transport the household uses most days; car, bus or train. It contains demographic information such as home ownership, income, and geographic information such as rural and urban residence. Most importantly for our question, the data provides a measure of how dense the rail network is in the location.

The section uses logit and probit models to estimate the demand for cars in "rail" cities and "non-rail" cities. It estimates the multinomial version of these models for "rail" cities. It then takes those parameter estimates to predict demand for rail in "non-rail" cities.

---

[20]https://nhts.ornl.gov/ or https://sites.google.com/view/microeconometricswithr/table-of-contents

5.7.1   National Household Travel Survey

The following code brings in the data and creates variables. It creates the "choice" variables for transportation use. Note that households may report using more than one mode, but the variables are defined exclusively. The code also adjusts variables for missing. The data uses a missing code of -9. This would be used as a value if it is not replaced with NA.

```
> x <- read.csv("hhpub.csv", as.is = TRUE)
> x$choice <- NA
> x$choice <- ifelse(x$CAR==1,"car",x$choice)
> x$choice <- ifelse(x$BUS==1,"bus",x$choice)
> x$choice <- ifelse(x$TRAIN==1,"train",x$choice)
> # Note that this overrules the previous choice.
> x$car1 <- x$choice=="car"
> x$train1 <- x$choice=="train"
> # adjusting variables to account for missing.
> x$home <- ifelse(x$HOMEOWN==1,1,NA)
> x$home <- ifelse(x$HOMEOWN>1,0,x$home)
> # home ownership
> x$income <- ifelse(x$HHFAMINC > 0, x$HHFAMINC, NA)
> # household income
> x$density <- ifelse(x$HTPPOPDN==-9,NA,x$HTPPOPDN)/1000
> # missing is -9.
> # population density
> # dividing by 1000 makes the reported results look nicer.
> x$urban1 <- x$URBAN==1  # urban versus rural
> y <- x[x$WRKCOUNT>0 & (x$MSACAT == 1 | x$MSACAT == 2),]
> # limit to households that may commute and those that
> # live in some type of city.
> y$rail <- y$RAIL == 1
> # an MSA with rail
> index_na <- is.na(rowSums(cbind(y$car1,y$train1,
+                                 y$home,y$HHSIZE,y$income,
+                                 y$urban1,y$density,y$MSACAT,
+                                 y$rail)))==0
> y <- y[index_na,] # drop missing
```

How different are "rail" cities from "non-rail" cities? The plan is to use demand estimates for cars, buses, and rail in cities with rail networks to predict demand for rail in other cities. However, cities with and without rail may differ in a variety of ways which may lead to different demand for rail between the two types of cities.

```
> vars <- c("car1","train1","home","HHSIZE","income",
+           "urban1","density")
```

```
> summ_tab <- matrix(NA,length(vars),2)
> for (i in 1:length(vars)) {
+    summ_tab[i,1] <- mean(y[y$rail==1,colnames(y)==vars[i]])
+    summ_tab[i,2] <- mean(y[y$rail==0,colnames(y)==vars[i]])
+ }
> row.names(summ_tab) <- vars
> colnames(summ_tab) <- c("Rail","No Rail")
```

	Rail	No Rail
car1	0.88	0.97
train1	0.10	0.01
home	0.73	0.75
HHSIZE	2.49	2.46
income	7.52	7.05
urban1	0.91	0.87
density	7.56	4.66

TABLE 5.2
Means by rail network density.

Table 5.2 presents summary results for each type of city. It shows that in cities with rail networks about 10% of the population uses trains most days, while it is only 1% of those in cities without dense rail networks. The different cities also differ in other ways including income and population density.

What would happen to the demand for rail if a city without one built it? Would demand increase to 10% as it is in cities with rail networks? Or would demand be different due to other differences between the cities?

5.7.2 Demand for Cars

We can also look at how the demand for cars varies between the two types of cities.

```
> # Without Rail
> y_nr <- y[y$rail == 0,]
> glm_nr <- glm(car1 ~ home + HHSIZE + income + urban1 +
+                density, data = y_nr,
+                family=binomial(link=logit))
> # With Rail
> y_r <- y[y$rail == 1,]
> glm_r <- glm(car1 ~ home + HHSIZE + income + urban1 +
+                density, data = y_r,
+                family=binomial(link=logit))
```

	Dependent variable:	
	car1	
	(1)	(2)
home	0.633***	0.498***
	(0.095)	(0.071)
HHSIZE	−0.018	0.050*
	(0.033)	(0.026)
income	0.133***	−0.068***
	(0.019)	(0.013)
urban1	−1.160***	−0.312*
	(0.245)	(0.187)
density	−0.055***	−0.109***
	(0.008)	(0.003)
Constant	3.775***	3.506***
	(0.270)	(0.211)
Observations	22,419	11,624
Note:	*p<0.1; **p<0.05; ***p<0.01	

TABLE 5.3
Logit estimates of demand for cars. Model (1) is demand in cities without a dense rail network. Model (2) is demand in cities with a rail network.

Table 5.3 shows that demand for cars varies given different access to rail. It is not clear how to interpret the coefficient estimates. There are lots of differences between the two types of cities, including the choices available to commuters. Nevertheless, we see the demand increasing in home ownership and decreasing in density.

5.7.3 Estimating Demand for Rail

We can set up the McFadden demand model for cars, buses and trains. The utility of car and train is relative to bus. Note that the value of train is assumed to be a function of density. The assumption is that trains have fixed station locations and in more dense cities, these locations are likely to be more easily accessible to the average person.

```
> X_r_car <- cbind(y_r$home, y_r$HHSIZE, y_r$income,
+                    y_r$urban1,0)
> X_r_train <- cbind(0,0,0,0,y_r$density)
> # train value is assumed to be determined by
> # population density.
> y1_r <- cbind(y_r$choice=="car",y_r$choice=="train")
```

The following presents the optimization procedure for the two multinomial choice models. Each uses the initial probit or logit model for starting values on the β term. To be clear, the estimation is done on cities with rail networks.

```
> par1 <- c(0,glm_r$coefficients)
> a1 <- optim(par=par1,fn=f_biprobit,y=y1_r,W1=X_r_car,
+              W2=X_r_train,K=100,control=list(trace=0,
+                                       maxit=10000))
> par2 <- glm_r$coefficients
> a2 <- optim(par=par2,fn=f_logit,y=y1_r,W1=X_r_car,
+              W2=X_r_train,control=list(trace=0,maxit=10000))
```

5.7.4 Predicting Demand for Rail

Once we estimate demand for car, bus, and rail in cities with rail networks, we can use the estimated parameters to predict demand for rail in a non-rail city. To do this, we combine the parameter estimates from rail cities with the observed characteristics of households in non-rail cities. This is done for both the multinomial probit and the multinomial logit.

```
> X_nr_car <- cbind(1,y_nr$home, y_nr$HHSIZE, y_nr$income,
+                    y_nr$urban1,0)
> X_nr_train <- cbind(1,0,0,0,0,y_nr$density)
> # Probit estimate
> set.seed(123456789)
```

```
> rho_t <- exp(a1$par[1])/(1 + exp(a1$par[1]))
> beta_t <- a1$par[2:length(a1$par)]
> W1 <- X_nr_car
> W2 <- X_nr_train
> W1b_t <- W1%*%beta_t
> W2b_t <- W2%*%beta_t
> K = 100
> u <- rnorm(K)
> p_00 <- p_10 <- rep(0, dim(W1)[1])
> for (u_k in u) {
+    u_k0 <- u_k < -W1b_t
+    p_00 <- p_00 +
+      u_k0*pnorm((-W2b_t - rho_t*u_k)/((1 - rho_t^2)^(.5)))
+    p_10 <- p_10 +
+      (1 - u_k0)*pnorm((W1b_t - W2b_t +
+                               (1 - rho_t)*u_k)/((1 - rho_t^2)^(.5)))
+ }
> p_train_pr <- (K - p_00 - p_10)/K
> mean(p_train_pr)

[1] 0.08872575

> # Logit estimate
> beta_hat <- a2$par
> W1b_hat <- W1%*%beta_hat
> W2b_hat <- W2%*%beta_hat
> p_train_lg <- exp(W2b_hat)/(1 + exp(W1b_hat) + exp(W2b_hat))
> mean(p_train_lg)

[1] 0.08313867
```

If a city without rail built one, the demand for rail would increase substantially. It would increase from 1% to about 9%. A substantial increase, almost to the 10% we see for cities with rail. Do you think this evidence is enough to warrant investment in rail infrastructure by smaller US cities? Can you create a bootstrap version of these estimators?

5.8   Discussion and Further Reading

The first part of the book presented methods that focused on the use of experimentation, or lack thereof. This part focuses on methods based on economic theory. This chapter discusses how the economic assumption of **revealed preference** can be used to identify the policy parameters of interest.

The Nobel Prize winning economist, Daniel McFadden, showed how combining economic theory and survey data could provide better predictions of demand for new products.

This chapter presented the standard discrete choice models of the logit and probit. The chapter introduces the maximum likelihood algorithm. It shows how this algorithm can be used to estimate OLS, logit and probit models.

The application of the logit and probit is to demand estimation. While these are standard methods, the reader should note that they are not really the way modern demand analysis is done. The chapter assumes that prices (or product characteristics) are exogenous. Modern methods use an IV approach to account for the endogeneity of prices. In particular, the ideas of Berry et al. (1995) have become standard in my field of industrial organization. Chapter 7 discusses this approach in more detail.

6

Estimating Selection Models

6.1 Introduction

Chapter 5 introduced the idea of using the economic assumption of **revealed preference** for estimating policy effects. Berkeley econometrician, Daniel Mc-Fadden, won the Nobel prize in economics for his work using revealed preference to estimate demand. McFadden was joined in the Nobel prize by University of Chicago econometrician, James Heckman. Heckman won for his work advocating the use of revealed preference to a broader range of problems.

Chapter 6 considers two related problems, **censoring** and **selection**. Censoring occurs when the value of a variable is limited due to some constraint. For example, we tend not to see wages below the federal minimum wage. The chapter shows our estimates can be biased when our statistical models expect the variable to go below the censored level. A standard method to account for censoring is to combine a probit with OLS. This combined model is called a **Tobit**. The chapter estimates a wage regression similar to Chapter 2's analysis of returns to schooling. The difference is that the regression accounts for censoring of wages at the minimum wage.

The **selection problem** is a generalization of the censoring problem. The data is censored due to some sort of "choice."[1] While McFadden considered problems where the choice was observed but the outcomes were not. Heckman examined a question where both the choice and the outcome of that choice are observed.

The chapter uses Heckman's model to analyze the gender-wage gap. The concern is that observed difference in wages by gender may underestimate the actual difference. Traditionally, many women did not have paid work because they have childcare or other uncompensated responsibilities. Whether or not a woman works full-time depends on the wage she is offered. We only observed the offers that were accepted, which means the offers are "selected." The Heckman model allows us to account for the "choice" of these women to work.

In addition to analyzing the gender wage gap, the chapter returns to the question of measuring returns to schooling. The chapter uses a version of the

---

[1] We will use the term "choice" to refer to assignment to treatment that is associated with some sort of economic decision. It does not mean that the observed individual actually had a "choice."

Heckman selection model to estimate the joint distribution of potential wages for attending college and not attending college.

6.2 Modeling Censored Data

Censoring refers to the issue that a variable is set to an arbitrary value such as 0. Say for example, that a variable must always have a positive value. When we look at hours worked, the values are all positive. The minimum number of hours a person can work is zero. Such restrictions on the values can make it difficult to use OLS and other methods described in the previous chapters.

The section presents the latent value model and the Tobit estimator.

6.2.1 A Model of Censored Data

Consider a model eerily similar to the model presented in the previous chapter. There is some latent outcome (y^*), where if this value is large enough we observe $y = y^*$. The variable y^* could be estimated with a standard OLS model. However, y^* is not fully observed in the data. Instead y is observed, where y is equal to y^* when that latent variable is above some threshold. Otherwise, the variable is equal to the threshold. Here the variable is censored at 0.

$$y_i^* = a + bx_i + v_i$$

$$y_i = \begin{cases} y_i^* \text{ if } y_i^* \geq 0 \\ 0 \text{ if } y_i^* < 0 \end{cases} \tag{6.1}$$

We can think of y_i^* as representing hours worked like in the example above.

6.2.2 Simulation of Censored Data

Consider a simulated version of the model presented above.

```
> set.seed(123456789)
> N <- 500
> a <- 2
> b <- -3
> x <- runif(N)
> u <- rnorm(N)
> y_star <- a + b*x + u
> y <- ifelse(y_star > 0, y_star, 0)
> lm1 <- lm(y ~ x)
> lm1$coefficients[2]
```

```
       x
-2.026422
```

Figure 6.1 shows that the relationship estimated with OLS is quite different from the true relationship. The true relationship has a slope of -3, while the estimated relationship is much flatter with a slope of -2. Can you see the problem?

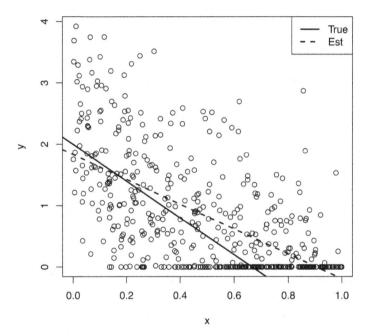

FIGURE 6.1
Plot of x and y with the relationship between y^* and x represented by the solid line. The estimated relationship between x and y is represented by the dashed line.

OLS does not provide a correct estimate of the relationship because the data is censored. Only positive values of y are observed in the data, but the true relationship implies there would in fact be negative values.

The implication is that our method of **averaging**, discussed in Chapter 1, no longer works. One solution is to limit the data so that it is not censored. Figure 6.1 suggests that for values of x below 0.6, the data is mostly not censored. This gives an **unbiased** estimate but at the cost of getting rid of more than half of the data. The term **biased** refers to whether we expect the estimated value to be equal to the true value. If it is biased, then we don't expect them to be equal. If it is **unbiased** we do expect them to be equal.[2] We generally use the term **efficient**, to refer to the likely variation in our estimate due to **sampling error**. Limiting the data makes our estimate less biased but less efficient.

```
> lm(y[x < 0.6] ~ x[x < 0.6])$coef[2]

x[x < 0.6]
 -2.993644

> length(y[x < 0.6])

[1] 291
```

In this case, limiting the sample to the data that is not censored leads to an estimate close to the true value of -3.

Another solution is to ignore the exact amount of positive values and estimate a probit. If we simplify the problem by setting all positive values to 1 we can use a standard probit. Again our estimate is also not efficient. We have thrown away a bunch of information about the value of y when it is not censored.

```
> glm(y > 0 ~ x, family = binomial(link = "probit"))$coef

(Intercept)           x
   2.015881    -3.013131
```

Again, despite throwing away information, the probit gives results that are pretty close to the true values of 2 and -3.

The solution presented below is to use a probit to account for the censoring and estimate OLS on the non-censored data. In particular, the Tobit is a maximum likelihood estimator that allows the two methods to be combined in a natural way. The estimator also uses all the information and so is more efficient than the solutions presented above.

6.2.3 Latent Value Model

One way to correct our estimate is to determine what the censored values of y^* are. At the very least, we need to determine the distribution of the latent values. Figure 6.2 presents the histogram of the observed values of y. While we observe the uncensored distribution of y^*, we have no idea what the censored distribution of y^* looks like. However, we may be willing to make an assumption about the shape of the distribution. In that case, it may be possible to estimate the distribution of the missing data using information from the data that is not censored.

The **latent value model** is very similar to the demand model presented in

---

[2]See Appendix A for a longer discussion of what it means for an estimator to be unbiased.

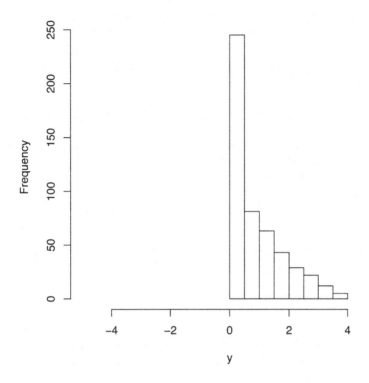

FIGURE 6.2
Histogram of the observed values of y.

the previous chapter. In both cases, there is some **latent** value that we are interested in measuring. We can write out the data generating process.

$$y_i^* = \mathbf{X}_i'\beta + v_i$$

$$y_i = \begin{cases} y_i^* \text{ if } y_i^* \geq 0 \\ 0 \text{ if } y_i^* < 0 \end{cases} \tag{6.2}$$

where y_i is the outcome of interest observed for individual i, y_i^* is the latent outcome of interest, $\mathbf{X}_i$ are observed characteristics that may both determine the latent outcome and the probability of the outcome being censored, β is a vector of parameters, and v_i represents the unobserved characteristics. In the demand model, the parameters represented each individual's preferences. In the case analyzed below, the latent outcome is the wage rate and the observed characteristics are the standard variables related to observed wages such as education, age and race.

If the unobserved characteristic of the individual is high enough, then the outcome is not **censored**. In that case, we have the OLS model. As shown in the previous chapter, if the unobserved term is distributed normally, $v_i \sim \mathcal{N}(0, \sigma^2)$, then we have the following likelihood of observing the data.

$$L(y_i, \mathbf{X}_i | y_i > 0, \beta, \sigma) = \frac{1}{\sigma} \phi(z_i) \tag{6.3}$$

where

$$z_i = \frac{y_i - \mathbf{X}_i'\beta}{\sigma} \tag{6.4}$$

where the **standard normal density** is denoted by ϕ and we need to remember that the **density** is a derivative, so there is an extra σ.

For the alternative case, we can use the **probit** model.

$$L(y_i, \mathbf{X}_i | y_i = 0, \beta) = \Phi(-\mathbf{X}_i'\beta) \tag{6.5}$$

The censored model with the normality assumption is called a **Tobit**. The great econometrician, Art Goldberger, named it for the great economist, James Tobin, and the great limited dependent variable model, the **probit** (Enami and Mullahy, 2009).[3]

6.2.4   Tobit Estimator

We can write out the **Tobit** estimator by combining the ideas from the **maximum likelihood** estimator of OLS and the probit presented in Chapter 5.

That is, find the parameters (β and σ) that maximize the probability that the model predicts the observed data.

$$\max{}_{\hat{\beta}, \hat{\sigma}} \quad \sum_{i=1}^{N} \mathbb{1}(y_i = 0) \log(\Phi(\hat{z}_i)) + \mathbb{1}(y_i > 0)(\log(\phi(\hat{z}_i)) - \log(\hat{\sigma})) \tag{6.6}$$

$$s.t. \quad \hat{z}_i = \frac{y_i - \mathbf{X}_i'\hat{\beta}}{\hat{\sigma}}$$

The notation $\mathbb{1}()$ is an **indicator** function. It is equal to 1 if what is inside the parentheses is true, 0 if false. As stated in Chapter 5, it is better for the computer to maximize the log likelihood rather than the likelihood. In addition, I made a slight change to the "probit" part which makes the description closer to the estimator in **R**.

Unlike the probit model we have an extra term for the distribution of the unobserved term (σ). This parameter is not identified in a discrete choice model but it is identified in a Tobit model. We need to be careful that we correctly include the effect of this parameter in the log likelihood function.

6.2.5   Tobit Estimator in R

We can use Equation (6.6) as pseudo-code for the estimator in **R**.

---

[3]According to former Tobin students, the estimator once took an army of graduate research assistants armed with calculators to estimate the model.

```
> f_Tobit <- function(par, y, X) {
+    X <- cbind(1,X)
+    sigma <- exp(par[1]) # use exp() to keep value positive.
+    beta <- par[2:length(par)]
+    is0 <- y == 0
+    # indicator function for y = 0.
+    z <- (y - X%*%beta)/sigma
+    log_lik <- -sum(is0*log(pnorm(z)) +
+                    (1 - is0)*(log(dnorm(z)) - log(sigma)))
+    # note the negative because we are minimizing.
+    return(log_lik)
+ }
> par1 <- c(0,lm1$coefficients)
> a1 <- optim(par=par1,fn=f_Tobit,y=y,X=x)
> # sigma
> exp(a1$par[1])

0.9980347

> # beta
> a1$par[2:length(a1$par)]

(Intercept)           x
   2.010674    -3.012778
```

The model estimates the three parameters pretty accurately, the shape parameter has a true value of 1, while $\hat{\beta}$ is pretty close to the true values of 2 and -3. To be clear, in the simulated data the unobserved term is in fact normally distributed, an important assumption of the Tobit model. What happens if you estimate this model with some other distribution?

6.3   Censoring Due to Minimum Wages

One of the standard questions in labor economics is determining the effect on earnings of an individual's characteristics, like education, age, race, and gender. For example, we may be interested in whether women are paid less than men for the same work. A concern is that our wage data may be censored. For example, in July 2009 the federal government increased the federal minimum wage to \$7.25. That is, it was generally illegal to pay people less than \$7.25.[4]

The section uses information on wages from 2010 and compares OLS to Tobit estimates.

---

[4]There are various exemptions such as for employees receiving tips.

6.3.1   National Longitudinal Survey of Youth 1997

The National Longitudinal Survey of Youth 1997 (NLSY97) is a popular data set for applied microeconometrics and labor economics. The data follows about 8,000 individuals across 18 years beginning in 1997. At the start of the data collection, the individuals are teenagers or in their early 20s.[5]

```
> x <- read.csv("NLSY97_min.csv", as.is = TRUE)
> x$wage <-
+    ifelse(x$CVC_HOURS_WK_YR_ALL.10 > 0 & x$YINC.1700 > 0,
+                    x$YINC.1700/x$CVC_HOURS_WK_YR_ALL.10,NA)
> x$wage <- as.numeric(x$wage)
> x$wage <- ifelse(x$wage < quantile(x$wage,0.90,
+                                    na.rm = TRUE), x$wage,NA)
> # topcode at the 90th percentile
> # this done to remove unreasonably high measures of wages.
> x$lwage <- ifelse(x$wage > 7.25, log(x$wage), 0)
> # note the 0s are used as an indicator in the Tobit function.
> x$ed <- ifelse(x$CV_HGC_EVER_EDT>0 &
+                    x$CV_HGC_EVER_EDT < 25,x$CV_HGC_EVER_EDT,NA)
> # removed very high values of education.
> x$exp <- 2010 - x$KEY.BDATE_Y - x$ed - 6
> x$exp2 <- (x$exp^2)/100
> # division makes the reported results nicer.
> x$female <- x$KEY.SEX==2
> x$black <- x$KEY.RACE_ETHNICITY==1
> index_na <- is.na(rowSums(cbind(x$lwage,x$wage,x$ed,x$exp,
+                                 x$black,x$female)))==0
> x1 <- x[index_na,]
```

To illustrate censoring we can look at wage rates for the individuals in NLSY97. Their **average wage rate** is calculated as their total income divided by the total number hours worked for the year. In this case, income and wages are measured in 2010. The code uses the censored variable where log wage is set to 0 unless the wage rate is above \$7.25 per hour.

Figure 6.3 gives some indication of the issue. We see that the distribution of wages is not symmetric and there seems to be higher than expected frequency just above the federal minimum wage. Actually, what is surprising is that there is a relatively large number of individuals with average wages below the minimum. It is unclear why that is, but it may be due to reporting errors or cases where the individual is not subject to the law.

---

[5]This data and other similar data sets are available from the Bureau of Labor Statistics here: https://www.nlsinfo.org/investigator/pages/login.jsp. This version can be downloaded from here: https://sites.google.com/view/microeconometricswithr/table-of-contents

```
> hist(x1[x1$wage<30,]$wage,xlab="wage rate",main="")
> abline(v=7.25, lwd=2)
```

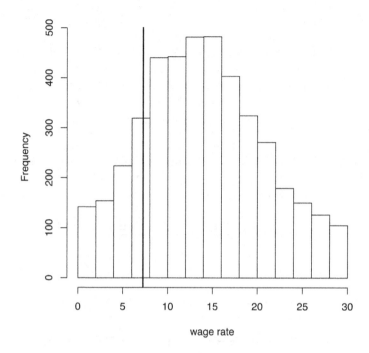

FIGURE 6.3
Histogram of the observed wages in 2010 from NLSY97. The vertical line is at
the minimum wage of $7.25.

6.3.2 Tobit Estimates

Even though it may provide inaccurate results, it is always useful to run OLS.
Here we use it as a comparison to see if the censoring affects our measurement of
how education, gender, and race affect wages. Following the argument presented
above, it is useful to also see the probit estimates. The probit accounts for
some of the impact of censoring although it throws away a lot of information.

```
> lm1 <- lm(lwage ~ ed + exp + exp2 + female + black, data=x1)
> glm1 <- glm(lwage > 0 ~ ed + exp + exp2 + female + black,
+                  data=x1, family=binomial(link="probit"))
```

Comparing OLS to probit and Tobit estimates we can see how the censoring
affects standard estimates of returns to experience, schooling, race and gender.

```
> par1 <- c(0,glm1$coefficients)
```

```
> a <- optim(par=par1,fn=f_Tobit,y=x1$lwage,
+              X=cbind(x1$ed,x1$exp,x1$exp2,x1$female,x1$black),
+              control = list(trace=0,maxit=10000))
> res_tab <- cbind(lm1$coefficients,glm1$coefficients,
+                    a$par[2:length(a$par)])
> res_tab <- rbind(res_tab,c(1,1,exp(a$par[1])))
> colnames(res_tab) <- c("OLS Est","Probit Est","Tobit Est")
> rownames(res_tab) <- c("intercept","ed","exp","exp sq",
+                          "female","black","sigma")
```

Table 6.1 presents a comparison of wage rate regressions between OLS, probit and the Tobit. Note that the values for σ are arbitrarily set to 1 for the OLS and probit. The Tobit estimates suggest that being female and being black lead to lower wages. The results suggest that censoring is leading to even greater effects than OLS would suggest.

	OLS Est	Probit Est	Tobit Est
intercept	-0.13	-1.51	-0.64
ed	0.14	0.14	0.16
exp	0.12	0.12	0.12
exp sq	-0.40	-0.42	-0.35
female	-0.17	-0.20	-0.19
black	-0.28	-0.31	-0.33
sigma	1.00	1.00	1.22

TABLE 6.1
OLS, probit and Tobit estimates on 2010 wage rates.

6.4   Modeling Selected Data

The Tobit model is about censoring. A close cousin of censoring is selection. In both cases we can think of the problem as having missing data. The difference is the reason for the missingness. In censoring, the data is missing because the outcome variable of interest is above (or below) some threshold. With selection, the data is missing because the individuals in the data have made a choice or have had some choice made for them.

Consider the problem of estimating returns to schooling for women. Compared to males, a large share of the female population don't earn wages. We have a selection problem if this choice is determined by how much these women would have earned if they had chosen to work. The observed distribution of wages for women may be systematically different than the unobserved distribution of wage *offers*. This difference may lead us to underestimate the gender wage gap.

6.4.1 A Selection Model

Consider a model similar to the one presented above. There is some latent outcome (y_i^*). This time, however $y_i = y_i^*$ if some other value z_i is large enough.

$$y_i^* = a + bx_i + v_{2i}$$

$$y_i = \begin{cases} y_i^* \text{ if } c + dz_i + v_{1i} \geq 0 \\ 0 \text{ if } c + dz_i + v_{1i} < 0 \end{cases} \tag{6.7}$$

where $\{v_1, v_2\} \sim \mathcal{N}(\mu, \Sigma)$, $\mu = \{0, 0\}$ and

$$\Sigma = \begin{bmatrix} 1 & \rho \\ \rho & 1 \end{bmatrix} \tag{6.8}$$

Selection is modeled as the correlation between the two unobserved terms v_1 and v_2, which is denoted ρ. The idea is that unobserved characteristics that determine the outcome of interest also determine whether the outcome is observed.[6]

We can think of y_i^* as representing hours worked, where the individual decides to work full-time based on factors such as child care cost or availability. Someone is more likely to work if they are offered a higher wage. That said, there are other characteristics such as the availability of child care that can also determine whether the person works.

Note if $z_i = x_i$, $a = c$, $b = d$ and $v_{1i} = v_{2i}$ then this is exactly the same model as the Tobit presented above.

6.4.2 Simulation of a Selection Model

Consider a simulated data set that is similar to the data created for the Tobit model. One difference is the z variable. Where previously, whether or not y was observed depended on x and v_2. Here it depends on z, v_1 and v_2. Importantly, neither z nor v_1 determine the observed value of y. In this data, the z variable is determining whether or not the y variable is censored.

```
> require(mvtnorm)
> set.seed(123456789)
> N <- 100
> a <- 6
> b <- -3
> c <- 4
```

---

[6]For simplicity, it is assumed that the variance is 1.

```
> d <- -5
> x <- runif(N)
> z <- runif(N)
> mu <- c(0,0)
> Sigma <- rbind(c(1,0.5),c(0.5,1))
> u <- rmvnorm(N, mean=mu, sigma=Sigma)
> # creates a matrix with two correlated random variables.
> y <- ifelse(c + d*z + u[,1] > 0, a + b*x + u[,2], 0)
> # OLS Model
> x1 <- x
> y1 <- y
> lm1 <- lm(y1 ~ x1)
> # OLS with subset of data (z < 0.6)
> x1 <- x[z < 0.6]
> y1 <- y[z < 0.6]
> lm2 <- lm(y1 ~ x1)
> # Probit
> y1 <- y > 0
> glm1 <- glm(y1 ~ z, family = binomial(link = "probit"))
```

Table 6.2 presents OLS and probit regression results for the simulated data. The outcome variable y is regressed on the two explanatory variables x and z. Note that in the simulation z only affects whether or not y is censored, while x affects the outcome itself. The results show that simply regressing y on x will give biased results. Model (1) is the standard OLS model with just the x variable. The estimate on the coefficient is far from the true value of -3. This is due to the censoring, similar to the argument made above. Again, we can restrict the data to observations less likely to be impacted by the censoring. Model (2) does this by restricting analysis to observations with small values of z. This restriction gives estimates that are pretty close to the true values. The probit model (Model (3)) accounts for the impact of z on y. The results from this model are pretty close to the true values.

6.4.3   Heckman Model

In algebra, the selection model is similar to the Tobit model.

$$y_i = \begin{cases} \mathbf{X}'_i\beta + \upsilon_{2i} \text{ if } \upsilon_{1i} \geq -\mathbf{Z}'_i\gamma \\ 0 \text{ if } \upsilon_{1i} < -\mathbf{Z}'_i\gamma \end{cases} \tag{6.9}$$

where $\{\upsilon_{1i}, \upsilon_{2i}\} \sim \mathcal{N}(\mu, \mathbf{\Sigma})$, $\mu = \{0,0\}$ and

$$\mathbf{\Sigma} = \begin{bmatrix} 1 & \rho \\ \rho & 1 \end{bmatrix} \tag{6.10}$$

In the Heckman model the "decision" to work or not, is dependent on a different set of observed and unobserved characteristics represented by $\mathbf{Z}$ and υ. Note

	Dependent variable:		
	y1		
	OLS		probit
	(1)	(2)	(3)
x1	−1.314*	−3.358***	
	(0.741)	(0.517)	
z			−4.905***
			(1.150)
Constant	4.397***	6.039***	4.025***
	(0.426)	(0.301)	(0.869)
Observations	100	64	100
R^2	0.031	0.405	
Note:			*p<0.1; **p<0.05; ***p<0.01

TABLE 6.2
Estimates on the simulated data where Model (1) is an OLS model, Model (2) is an OLS model on data where $z < 0.6$ and Model (3) is a probit. The true values for Models (1) and (2) are -3 and 6 (going down), while for Model (3) the true values are -5 and 4.

that these could be exactly the same as the Tobit. It may be that $\mathbf{X} = \mathbf{Z}$ and $\rho = 1$. That is, the Heckman model is a generalization of the Tobit model. In the example of women in the workforce, the Heckman model allows things like childcare costs to determine whether women are observed earning wages. It also allows unobserved characteristics that affect the probability of being in the workforce also affect the wages earned.

6.4.4   Heckman Estimator

As with the Tobit model, we can use maximum likelihood to estimate the model. The likelihood of observing the censored value $y_i = 0$ is determined by a probit.

$$L(y_i, \mathbf{Z}_i | y_i = 0, \gamma) = \Phi(-\mathbf{Z}_i'\gamma) \qquad (6.11)$$

The likelihood of observing a censored value of y is determined by the Zs and the vector γ. Note again, the variance of the normal distribution is not identified in the probit model and so is set to 1.

The likelihood of a strictly positive value of y_i is more complicated than for the censored value. If the unobserved terms determining whether y is censored (v_1) and its value (v_2) are independent then this likelihood would be as for the Tobit. However, the two unobserved terms are not independent. This means that we need to condition on v_1 to determine the likelihood of v_2.

Unfortunately, it is rather gruesome to write down the likelihood in this way. Therefore, we take advantage of Bayes's rule and write down the likelihood of v_1 conditional on the value of v_2.[7]

$$L(y_i, \mathbf{X}_i, \mathbf{Z}_i | y_i > 0, \beta, \gamma, \rho) = \left(1 - \Phi\left(\frac{-\mathbf{Z}_i'\gamma - \rho v_{2i}}{\sqrt{(1-\rho^2)}}\right)\right)\phi(v_{2i}) \qquad (6.12)$$

where $v_{2i} = y_i - \mathbf{X}_i'\beta$. Note how the conditioning is done in order to keep using the standard normal distribution.[8]

We want to find the parameters (β, γ and ρ) that maximize the probability that the model predicts the observed data.

$$\max_{\hat{\beta}, \hat{\gamma}, \hat{\rho}} \quad \sum_{i=1}^{N} \mathbb{1}(y_i = 0) \log\left(\Phi\left(-\mathbf{Z}_i'\hat{\gamma}\right)\right)$$
$$+ \mathbb{1}(y_i > 0)\left(\log\left(\Phi\left(\frac{\mathbf{Z}_i'\hat{\gamma} + \hat{\rho}(y_i - \mathbf{X}_i'\hat{\beta})}{\sqrt{(1-\hat{\rho}^2)}}\right)\right)\right) \qquad (6.13)$$
$$+ \log\left(\phi(y_i - \mathbf{X}_i'\hat{\beta})\right)\right)$$

---

[7]When dealing with joint distributions it is useful to remember the relationship $\Pr(A, B) = \Pr(A|B)\Pr(B) = \Pr(B|A)\Pr(A)$ where A and B represent events associated with each of two random variables. The term on the left-hand side represents the joint distribution. The middle term is the probability of observing A conditional on the value of B, multiplied by the probability of observing B. The term on the far right is the other way around.

[8]I can never remember exactly how to do this, so I always keep a copy of Goldberger (1991) close by. In **R** you can sometimes get around remembering all this and use a package like `mvtnorm` to account for multivariate normal distributions.

Note that I have taken advantage of the fact that normals are symmetric.

To make this presentation a little less messy, I assumed that the distribution of unobserved terms is a **bivariate standard normal**. That is, the variance terms are 1. As with the probit, the variance of v_1 is not identified, but as with the Tobit, the variance of v_2 is identified.[9] Nevertheless, here it is assumed to be 1.

6.4.5   Heckman Estimator in R

The code for the Heckman estimator is very similar to the code for the Tobit estimator. The difference is that this estimator allows for a set of characteristics that determine whether or not the outcome variable is censored.

```
> f_Heckman <- function(par,y, X_in, Z_in = X_in) {
+    # defaults to Z_in = X_in
+    X <- cbind(1,X_in)
+    Z <- cbind(1,Z_in)
+    is0 <- y == 0 # indicator function
+    rho <- exp(par[1])/(1 + exp(par[1]))
+    # this is the sigmoid function
+    # Note that in actual fact rho is between -1 and 1.
+    beta <- par[2:(1+dim(X)[2])]
+    gamma <- par[(2 + dim(X)[2]):length(par)]
+    Xb <- X%*%beta
+    Zg <- Z%*%gamma
+    Zg_adj <- (Zg + rho*(y - Xb))/((1 - rho^2)^(.5))
+    log_lik <- is0*log(pnorm(-Zg)) +
+        (1 - is0)*(log(pnorm(Zg_adj)) +
+                        log(dnorm(y - Xb)))
+    return(-sum(log_lik))
+ }
> par1 <- c(0,lm1$coefficients,glm1$coefficients)
> a1 <- optim(par=par1,fn=f_Heckman,y=y,X=x,Z=z)
> # rho
> exp(a1$par[1])/(1 + exp(a1$par[1]))

0.5357194

> # beta
> a1$par[2:3]

(Intercept)          x1
   5.831804    -2.656908

> # gamma
> a1$par[4:5]
```

---

[9]See discussion of the bivariate probit in the previous chapter.

```
(Intercept)              z
   4.406743    -5.488539
```

The Heckman estimator does a pretty good job of estimating the true parameters. The true $\beta = \{6, -3\}$, while the true $\gamma = \{4, -5\}$ and $\rho = 0.5$.

6.5 Analyzing the Gender Wage Gap

We can analyze the difference in wages between men and women using the NLSY97. Here we use the data from 2007 in order to minimize issues due to censoring. The analysis is also limited to "full-time" workers, those working more than an average of 35 hours per week.[10] A Heckman model is used to adjust for selection.

6.5.1 NLSY97 Data

The data is from NLSY97 with hours and income from 2007.[11] The analysis is limited to individuals working more than 1750 hours per year.

```
> x <- read.csv("NLSY97_gender_book.csv")
> x$wage <-
+   ifelse(x$CVC_HOURS_WK_YR_ALL.07_XRND > 0,
+          x$YINC.1700_2007/x$CVC_HOURS_WK_YR_ALL.07_XRND, 0)
> x$lwage <- ifelse(x$wage > 1, log(x$wage), 0)
> x$fulltime <- x$CVC_HOURS_WK_YR_ALL.07_XRND > 1750
> x$lftwage <- ifelse(x$lwage > 0 & x$fulltime, x$lwage, 0)
> x$female <- x$KEY_SEX_1997==2
> x$black <- x$KEY_RACE_ETHNICITY_1997==1
> x$age <- 2007 - x$KEY_BDATE_Y_1997
> x$age2 <- x$age^2
> x$college <-  x$CV_HIGHEST_DEGREE_0708_2007 >= 3
> x$south <- x$CV_CENSUS_REGION_2007==3
> x$urban <- x$CV_URBAN.RURAL_2007==1
> x$msa <- x$CV_MSA_2007 > 1 & x$CV_MSA_2007 < 5
> x$married <- x$CV_MARSTAT_COLLAPSED_2007==2
> x$children <- x$CV_BIO_CHILD_HH_2007 > 0
> index_na <- is.na(rowSums(cbind(x$black,x$lftwage,x$age,
+                                 x$msa, x$urban,x$south,
```

---

[10] Assuming a 50 week work-year.

[11] Again, this data set is available from the Bureau of Labor Statistics here: https://www.nlsinfo.org/investigator/pages/login.jsp. This version can be downloaded from here: https://sites.google.com/view/microeconometricswithr/table-of-contents

```
+                                      x$college,x$female,
+                                      x$married,x$children)))==0
> x1 <- x[index_na,]
> x1_f <- x1[x1$female,]
> x1_m <- x1[!x1$female,]
> # split by gender
```

Figure 6.4 presents the log densities of wage rates for male and female full-time workers. It shows that female wages are shifted down. The question from the analysis above is whether the true difference is much larger. Is the estimated distribution of wages for females biased due to selection? Asked another way, is the distribution of female wages shifted up relative to the distribution of female wage *offers*?

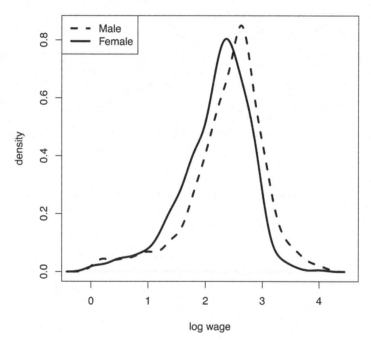

FIGURE 6.4
Density of full-time log wages by gender.

Part of the explanation for the difference may be differences in education level, experience or location. We can include these additional variables in the analysis.

The regressions presented in Table 6.3 suggest that there is a substantial female wage gap. This is shown in two ways, first by comparing Model (1) to Model (2). These are identical regressions except that Model (1) is just on male

wage earners and Model (2) is just on female wage earners. The regressions are
similar except for the intercept term. Model (2) is substantially shifted down
relative to Model (1). The second way is by simply adding a dummy for female
in Model (3). The negative coefficient on the dummy also suggests that there is
a substantial gender wage gap, even accounting for education and experience.

```
> lm1 <- lm(lftwage ~ age + age2 + black + college +
+              south + msa, data=x1_m)
> lm2 <- lm(lftwage ~ age + age2 + black + college +
+              south + msa, data=x1_f)
> lm3 <- lm(lftwage ~ female + age + age2 + black + college +
+              south + msa, data=x1)
```

6.5.2   Choosing to Work

While the previous analysis suggests a substantial gender wage gap, that gap
may be underestimated. This would occur if the women observed in the work
force were the ones more likely to have received higher wage offers. The first
step to estimate the distribution of wage offers is to estimate the "choice to
work." I put choice in quotations because I am not assuming that all women
are actually making a choice. I am assuming that whether a woman works or
not depends in part on what the woman expects to earn.

```
> glm1 <- glm(lftwage > 0 ~ college, data = x1_f)
> glm2 <- glm(lftwage > 0 ~ college + south, data = x1_f)
> glm3 <- glm(lftwage > 0 ~ college + south + married +
+                children, data = x1_f)
```

Table 6.4 shows that having a college education substantially increases
the likelihood that a woman will work. The analysis also suggests that being
married and having children affects the likelihood of working although the
coefficient on married is not statistically significantly different from zero and
the coefficient on children is positive.

6.5.3   Heckman Estimates of Gender Gap

In order to use the Heckman model, the outcome variable is normalized. The
model assumes that income for men and women is determined by experience,
education and location. For men it is assumed that if we don't observe a man
working full-time, it is something idiosyncratic about the man. In contrast, we
assume that women are **selected** to work full-time based on education, location,
number of children and whether they are married or not.

```
> x1$lftwage_norm <-
+   (x1$lftwage-mean(x1$lftwage))/sd(x1$lftwage)
```

	Dependent variable:		
	lftwage		
	(1)	(2)	(3)
female			−0.660***
			(0.045)
age	0.990	0.430	0.292
	(1.148)	(0.807)	(0.672)
age2	−0.017	−0.008	−0.004
	(0.023)	(0.016)	(0.013)
black	−0.660***	0.020	−0.255***
	(0.083)	(0.057)	(0.048)
college	0.547***	0.620***	0.596***
	(0.133)	(0.082)	(0.072)
south	0.053	−0.025	−0.012
	(0.080)	(0.055)	(0.047)
msa	0.030	0.038	0.048
	(0.143)	(0.120)	(0.093)
Constant	−12.470	−5.341	−3.297
	(14.382)	(10.084)	(8.412)
Observations	1,076	1,555	2,631
R^2	0.097	0.042	0.115
Note:		*p<0.1; **p<0.05; ***p<0.01	

TABLE 6.3
OLS estimates of log wages for full-time workers in 2007 from NLSY97. Model (1) is for male workers. Model (2) is for female workers. Model (3) includes both genders but a dummy variable for female.

	Dependent variable:		
	lftwage >0		
	(1)	(2)	(3)
college	0.219***	0.218***	0.215***
	(0.036)	(0.036)	(0.036)
south		0.014	0.015
		(0.023)	(0.023)
married			−0.037
			(0.089)
children			0.126**
			(0.062)
Constant	0.284***	0.278***	0.157**
	(0.012)	(0.016)	(0.062)
Observations	1,555	1,555	1,555
Note:	*p<0.1; **p<0.05; ***p<0.01		

TABLE 6.4

Probit estimates of the "choice" to work for females in NLSY97.

```
> x1$lftwage_norm3 <- ifelse(x1$lftwage==0,0,x1$lftwage_norm)
> y <- x1$lftwage_norm3
> X <- cbind(x1$age,x1$age2,x1$black,x1$college,
+             x1$south,x1$msa)
> Z <- cbind(x1$college,x1$south,x1$married,x1$children)
> y_f <- y[x1$female]
> X_f <- X[x1$female,]
> Z_f <- Z[x1$female,]
> par1 <- c(0,lm2$coefficients,glm3$coefficients)
> a <- optim(par=par1,fn=f_Heckman,y=y_f,X=X_f,Z=Z_f,
+             control = list(trace=0,maxit=10000))
> beta_hat <- a$par[2:8]
> y_adj <- cbind(1,X_f)%*%beta_hat + rnorm(dim(X_f)[1])
> X_m <- X[!x1$female,]
> y_adj_m <- cbind(1,X_m)%*%beta_hat + rnorm(dim(X_m)[1])
```

Figure 6.5 presents density estimates after accounting for selection into full-time work. It shows that the distribution of female wage offers is shifted much further down than the standard estimate. In order to account for observed differences between men and women, the chart presents a density of wage for women but with their observed characteristics set to the same values as men. Accounting for other observed differences between men and women has little effect.

6.6 Back to School Returns

This section returns to the question of whether an additional year of schooling increases income. This approach is similar to the IV approach presented in Chapter 3. It shows that the effect of college is heterogeneous and not always positive.

6.6.1 NLSM Data

We can use the NLSM data used by Card (1995) to compare the Heckman model with the IV approach.[12] In this analysis, a person is assumed to have gone to college if they have more than 12 years of education.

```
> x <- read.csv("nls.csv",as.is=TRUE)
> x$lwage76 <- as.numeric(x$lwage76)
> x1 <- x[is.na(x$lwage76)==0,]
> x1$lwage76_norm <-
```

---

[12]See a discussion of this data in Chapters 1 and 2.

```
> plot(density(y_adj, na.rm = TRUE),type = "l",lwd=3,
+       xlab="log wage",ylab="density",main="",ylim=c(0,1))
> # ylim sets the limits of the y-axis.
> lines(density(y_adj_m, na.rm = TRUE),lwd=3,lty=2)
> lines(density(x1[x1$female & x1$lftwage > 0,]$lftwage_norm,
+                 na.rm = TRUE), lty=3,lwd=3)
> lines(density(x1[!x1$female & x1$lftwage > 0,]$lftwage_norm,
+                 na.rm = TRUE), lwd=3,lty=4)
> legend("topleft",c("F (Heck est)","F (w/M chars)",
+                     "F (Std est)","M (Std est)"),
+        lwd=3,lty=c(1:4))
```

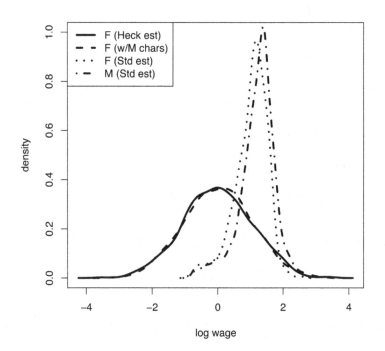

FIGURE 6.5
Density of wages by gender. Selection adjusted estimate is a solid line. Adjusted
for male characteristics is the dashed line.

```
+   (x1$lwage76 - mean(x1$lwage76))/sd(x1$lwage76)
> # norm log wages for Heckman model
> x1$exp <- x1$age76 - x1$ed76 - 6 # working years after school
> x1$exp2 <- (x1$exp^2)/100
> x1$college <- x1$ed76 > 12
```

6.6.2 College vs. No College

We can compare the differences in income for people who attended college and those that did not.

```
> lm_c <- lm(lwage76_norm ~ exp + exp2 + black + reg76r +
+               smsa76r, data=x1[x1$college,])
> lm_nc <- lm(lwage76_norm ~ exp + exp2 + black + reg76r +
+               smsa76r, data=x1[!x1$college,])
```

	Dependent variable:	
	lwage76_norm	
	(1)	(2)
exp	0.175***	0.182***
	(0.028)	(0.036)
exp2	−0.603***	−0.628***
	(0.184)	(0.149)
black	−0.372***	−0.565***
	(0.067)	(0.051)
reg76r	−0.219***	−0.469***
	(0.050)	(0.050)
smsa76r	0.379***	0.396***
	(0.056)	(0.048)
Constant	−0.785***	−1.231***
	(0.108)	(0.209)
Observations	1,521	1,489
R^2	0.152	0.248
Note:	*$p<0.1$; **$p<0.05$; ***$p<0.01$	

TABLE 6.5
OLS estimates of normalized log wages for males in the NLSM data. Model (1) is for males who attended college. Model (2) is for males who did not attend college.

Table 6.5 presents the results from OLS estimates for two groups: males that attend college and males that do not attend college. Note that this model is a little different from results presented in earlier chapters because I have normalized log wages for use in the Heckman model. Looking at the intercept term we see that the distribution of wages for males who attend college shifted up relative to the distribution of wages of males who do not.

6.6.3   Choosing College

Again we don't mean that a person is literally choosing whether or not to attend college. We mean that there are unobserved characteristics of the individual that are related to both attending college and earning income once graduated.

```
> # Probit Estimate
> glm1 <- glm(college ~ nearc4,
+             family = binomial(link = "probit"), data=x1)
> glm2 <- glm(college ~ nearc4 + momdad14,
+             family = binomial(link = "probit"), data=x1)
> glm3 <- glm(college ~ nearc4 + momdad14 + black + smsa66r,
+           family = binomial(link = "probit"),data=x1)
> glm4 <- glm(college ~ nearc4 + momdad14 + black + smsa66r
+             + reg662 + reg663 + reg664 + reg665 + reg666 +
+             reg667 + reg668 + reg669,
+             family = binomial(link = "probit"), data=x1)
```

Table 6.6 reiterates results we have seen in earlier chapters. Growing up near a 4-year college is associated with a higher probability of attending college; so is growing up with both parents. It is also related to race and where the person was living in 1966. Men living in cities were more likely to attend college.

6.6.4   Heckman Estimates of Returns to Schooling

The Heckman estimator is very similar to the IV estimator presented in Chapter 3.[13] I have highlighted the similarity by using the Z notation. In both cases, we have a policy variable X (college attendance) that is determined by instrument-like variables Z (distance to college, location in 1966, and whether they had both parents at home at 14). Formally, there is no requirement about the exogeneity of the Z variables. That is, there may be unobserved characteristics that determine both Z and X. The reason is that the parameters can be determined given the parametric assumptions we have made. That said, many researchers prefer to rely on the exogeneity relative to relying on the parametric assumptions. In some sense, this model is more general than the IV model. It allows for multiple instruments, which the IV does, but only if using GMM as shown in Chapter 8.

A probit is used to determine whether the person attends college or not.[14]

---

[13] I have simplified things somewhat in order to use the Heckman model presented above.

[14] This is often called a **control function**, and the approach is called a **control function approach**.

| | Dependent variable: | | | |
| | college | | | |
	(1)	(2)	(3)	(4)
nearc4	0.307***	0.316***	0.223***	0.237***
	(0.049)	(0.050)	(0.056)	(0.058)
momdad14		0.552***	0.411***	0.420***
		(0.058)	(0.060)	(0.061)
black			−0.502***	−0.466***
			(0.059)	(0.064)
smsa66r			0.139**	0.130**
			(0.055)	(0.057)
Observations	3,010	3,010	3,010	3,010
Note:				$^{*}p<0.1$; $^{**}p<0.05$; $^{***}p<0.01$

TABLE 6.6
Probit estimates of the "choice" to attend college for males in the NLSM data.
Model (4) includes regional dummies for 1966 (not reported).

The probit uses some of the instrumental variables discussed in Chapter 3 such as distance to college and whether the person lived with their parents at 14. Notice that these variables are not included in the variables determining income. Also note that the college function does not include variables that will affect income in 1976. The assumption is that the college decision had to do with factors that were true for the person in 1966, while income in 1976 has to do with factors that are true for the person in 1976.

```
> X <- cbind(x1$exp,x1$exp2,x1$black,x1$reg76r,x1$smsa76r)
> Z <- cbind(x1$nearc4,x1$momdad14,x1$black,x1$smsa66r,
+            x1$reg662,x1$reg663,x1$reg664,x1$reg665,
+            x1$reg666,x1$reg667,x1$reg668,x1$reg669)
> # College
> y_c <- x1[x1$college,]$lwage76_norm
> X_c <- X[x1$college,]
> Z_c <- Z[x1$college,]
> par1 <- c(0,lm_c$coefficients,glm4$coefficients)
> a_c <- optim(par=par1,fn=f_Heckman,y=y_c,X=X_c,Z=Z_c,
+            control = list(trace=0,maxit=100000))
> # No college
> y_nc <- -x1[!x1$college,]$lwage76_norm
> # negative in order to account for correlation in income
> # and no college.
> X_nc <- X[!x1$college,]
> Z_nc <- Z[!x1$college,]
> par1 <- c(0,lm_nc$coefficients,-glm4$coefficients)
> a_nc <- optim(par=par1,fn=f_Heckman,y=y_nc,X=X_nc,Z=Z_nc,
+            control = list(trace=0,maxit=100000))
> # Predicted wages
> mu <- c(0,0)
> rho_hat <- exp(a_c$par[1])/(1 + exp(a_c$par[1]))
> Sigma_hat <- cbind(c(1,rho_hat),c(rho_hat,1))
> u <- rmvnorm(dim(x1)[1],mu,Sigma_hat)
> beta_c_hat <- a_c$par[2:(length(lm_c$coefficients)+1)]
> beta_nc_hat <- a_nc$par[2:(length(lm_nc$coefficients)+1)]
> x1$coll_y <- cbind(1,X)%*%beta_c_hat + u[,1]
> x1$nocoll_y <- -cbind(1,X)%*%beta_nc_hat + u[,2]
```

This is not a "full" Heckman selection model. I have estimated two separate models, one on choosing college and the other on not choosing college. Note that I have used negatives to account for the "not" decision. This is done in order to simplify the exposition. But there are some costs, including the fact that there are two different estimates of the correlation term. In the results presented in Figure 6.6, the correlation coefficient from the first model is used. Can you write down and estimate a full model?

6.6.5 Effect of College

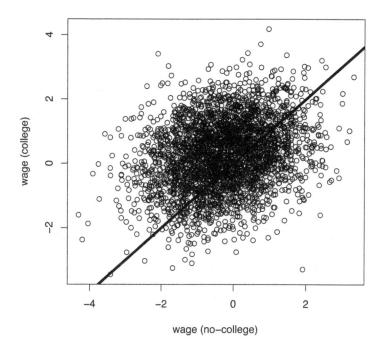

FIGURE 6.6
Plot of predicted potential wages by college and no-college. The 45 degree line is presented.

The Heckman model suggests that the effect of college is heterogeneous. While most benefit from college, not all benefit. Figure 6.6 presents a plot of the predicted wage for each person in the sample, for both the case where they went to college and the case where they did not go to college. For those sitting above the 45 degree line, their predicted wage is higher going to college than not going to college. For those below the 45 degree line, it is the opposite. Their predicted wages are higher from not attending college. Given all the assumptions, it is not clear exactly what we should take from this, but it is interesting that so much weight is above the 45 degree line. Remember that only 50% of the data actually went to college. The model assumes that people are "choosing" not to go to college based on beliefs about what they would earn if they did. The model is estimated with two Heckman regressions rather than a full "switching" model which would allow for a richer error structure.

We can compare this approach to the estimates of the ATE and LATE using OLS and IV. The estimated average treatment effect is 0.136, which is substantially higher than the OLS estimate of 0.075. It is similar to the IV and

LATE estimates. This should not be too surprising given how closely related the Heckman model is to the IV model.

```
> mean(x1$coll_y - x1$nocoll_y)/4
```

```
[1] 0.1357854
```

6.7   Discussion and Further Reading

The chapter introduces the **Tobit model** which can be used to estimate data that is **censored**. It also introduces the **Heckman selection model**, which is used to estimate the gender-wage gap and returns to schooling. When using limited dependent variable models, including the ones discussed in this chapter and the previous chapter, I often go back to appropriate chapters of Greene (2000).

This chapter showed how the economic assumption of revealed preference can be used for a broader range of problems than just demand estimation. In particular, in labor economics we often have data in which we observe both a decision and the outcome of the decision. The **Roy model** considers the situation where an individual is choosing between two employment sectors. We observe the choice and the wages in the chosen sector. Heckman and Honore (1990) show that with enough exogenous variation in wages and revealed preference, the joint distribution of wages across both sectors is identified. Recent work considers this model under even weaker assumptions (Henry et al., 2020).

The chapter re-analyzes the returns to schooling data used in Card (1995). It uses a version of the Heckman model to find average returns similar to the IV estimates. The model also estimates the joint distribution of returns and shows that for some individuals the expected income from attending college may be *less* than the expected income from not attending college.

7

Demand Estimation with IV

7.1 Introduction

When I began teaching microeconometrics a few years ago I read up on the textbook treatment of demand estimation. There was a lot of discussion about estimating logits and about McFadden's model. It looked a lot like how I was taught demand estimation 25 years before. It looked nothing like the demand estimation that I have been doing for the last 20 years. Demand estimation is integral to antitrust analysis. It is an important part of marketing and business strategy. But little of it actually involves estimating a logit. Modern demand estimation combines the insights of instrumental variable estimation and game theory.

My field, industrial organization, changed dramatically in the 1970s and 1980s as game theory became the major tool of analysis. A field that had been dominated by industry studies quickly became the center of economic theory and the development of game theoretic analysis in economics. By the time I got to grad school in the mid 1990s, the field was changing again. New people were looking to combine game theory with empirical analysis. People like Susan Athey, Harry Paarsh and Phil Haile started using game theory to analyze timber auctions and oil auctions. Others like Steve Berry and Ariel Pakes were taking game theory to demand estimation.

Game theory allows the econometrician to make inferences from the data by theoretically accounting for the way individuals make decisions and interact with each other. As with discrete choice models and selection models, we can use economic theory to help uncover unobserved characteristics. Again, we assume that economic agents are optimizing. The difference here is that we allow economic agents to explicitly interact and we attempt to model that interaction. Accounting for such interactions may be important when the number of agents is small and it is reasonable to believe that these individuals do in fact account for each other's actions.

This chapter presents a standard model of competition, the Hotelling model. It presents two IV methods for estimating demand from simulated data generated by the Hotelling model. It introduces the idea of using both cost shifters and demand shifters as instruments. The chapter takes these tools to the question of estimating the value of Apple Cinnamon Cheerios.

7.2   Modeling Competition

In the late 1920s, the economist and statistician, Harold Hotelling, developed a model of how firms compete. Hotelling wasn't interested in "perfect competition," and its assumption of many firms competing in a market with homogeneous products. Hotelling was interested in what happened when the number of firms is small and the products are similar but not the same. Hotelling (1929) was responding to an analysis written by the French mathematician, Joseph Bertrand, some 80 years earlier. Bertrand, in turn, was responding to another French mathematician, Antoine Cournot, whose initial analysis was published in the 1830s.

All three were interested in what happens when two firms compete. In Cournot's model, the two firms make homogeneous products. They choose how much to produce and then the market determines the price. Cournot showed that this model leads to much higher prices than was predicted by the *standard* (at the time) model of competition. Bertrand wasn't convinced. Bertrand considers the same case but had the firms choose prices instead.

Imagine two hotdog stands next to each other. They both charge $2.00 a hotdog. Then one day, the left stand decides to charge $1.90 a hotdog. What do you think will happen? When people see that the left stand is charging $1.90 and the right one is charging $2.00, they are likely to buy from the left one. Seeing this, the right hotdog stand reacts and sets her price at $1.80 a hotdog. Seeing the change, everyone switches to the right stand. Bertrand argued that this process will lead prices to be bid down to marginal cost. That is, with two firms the model predicts that prices will be the *same* as the standard model.

Hotelling agreed that modeling the firms as choosing price seemed reasonable, but was unconvinced by Bertrand's argument. Hotelling suggested a slight change to the model. Instead of the two hotdog stands being next to each other, he placed them at each end of the street. Hotelling pointed out that in this case even if the left stand was 10c cheaper, not everyone would switch away from the right stand. Some people have their office closer to the right stand and are unwilling to walk to the end of the street just to save a few cents on their hotdog. Hotelling showed that in this model prices were again much higher than for the standard model.

When I think about competition, it is Hotelling's model that I have in my head.

7.2.1   Competition is a Game

Hotelling, Cournot and Bertrand all modeled competition as **a game**. **A game** is a formal mathematical object which has three parts; **players**, **strategies** and **payoffs**. In the game considered here, the players are the two firms. The strategies are the actions that the players can take given the information available to them. The payoffs are the profits that the firms make. Note that in Cournot's game, the strategy is the quantity that the firm chooses to sell. In Bertrand's game, it is the price that the firm chooses to sell at. Cournot's model is a reasonable representation of an exchange or auction. Firms decide how much to put on the exchange and prices are determined by the exchange's mechanism. In Bertrand's game, the firm's post prices and customers decide how much to purchase.

Consider the following pricing game represented in Table 7.1. There are two firms, Firm 1 and Firm 2. Each firm chooses a price, either p_L or p_H. The firm's profits are in the cells. The first number refers to the profit that Firm 1 receives.

$\{F_1, F_2\}$	p_{2L}	p_{2H}
p_{1L}	{1,1}	{3,0}
p_{1H}	{0,3}	{2,2}

TABLE 7.1
A pricing game where Firms 1 and 2 choose prices $\{p_{jL}, p_{jH}\}$.

What do you think will be the outcome of the game? At which prices do both firms make the most money? If both firms choose p_H the total profits will be 4, which is split evenly. Will this be the outcome of the game?

No. At least it won't be the outcome if the outcome is a **Nash equilibrium**. The outcomes of the games described by Hotelling, Bertrand and Cournot are all **Nash equilibrium**. Interestingly, John Nash, didn't describe the equilibrium until many years later, over 100 years later in the case of Bertrand and Cournot. Even the definition of **a game** didn't come into existence until the work of mathematicians like John von Neumann in the early 20th century.

A **Nash equilibrium** is where each player's strategy is optimal given the strategies chosen by the other players. Here, a Nash equilibrium is where Firm 1's price is optimal given Firm 2's price, *and* Firm 2's price is optimal given Firm 1's price. It is not a Nash equilibrium for both firms to choose p_H. If Firm 2 chooses p_{2H}, then Firm 1 should choose p_{1L} because the payoff is 3, which is greater than 2. Check this by looking at the second column of the table and the payoffs for Firm 1, which are the first element of each cell.

The Nash equilibrium is $\{p_{1L}, p_{2L}\}$. You can see that if Firm 2 chooses p_{2L}, then Firm 1 earns 1 from also choosing p_{1L} and 0 from choosing p_{2H}. The same argument can be made for Firm 2 when Firm 1 chooses p_{1L}. Do you think this outcome seems reasonable?

7.2.2 Hotelling's Line

Figure 7.1 represents Hotelling's game. There are two firms L and R. Customers for the two firms "live" along the line. Customers prefer to go to the closer firm

L x_L R

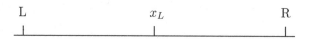

FIGURE 7.1
Hotelling's line with Firm L located at 0 and Firm R located at 1. Everyone "living" left of x_L purchases from Firm L.

if the products and prices are otherwise the same. Hotelling's key insight is that while firms often compete by selling similar products, these products may not be identical. Moreover, some people may prefer one product to the other. Some people actually prefer Pepsi to Coke or 7Up to Sprite. The location on the line represents how much certain customers prefer L to R.

Let Firm L lie at 0 and Firm R lie at 1. An infinite number of customers are located between 0 and 1. This infinite number of customers has a mass of 1. That is, we can think of location as a probability of purchasing from Firm L. Consider the utility of a customer at location $x \in [0, 1]$. This customer buys from L if and only if the following inequality holds.

$$U_L = v_L - x - \beta p_L > v_R - (1 - x) - \beta p_R = U_R \tag{7.1}$$

where v_L and v_R are the value of L and R respectively, p_L and p_R are the prices and β represents how customers relate the product to prices. The customer located at x_L is indifferent between buying from L or R.

$$v - x_L - \beta(p_L - p_R) = -1 + x_L$$

$$x_L = \frac{v + 1 - \beta p_L + \beta p_R}{2} \tag{7.2}$$

where $v = v_L - v_R$. This is also the demand for Firm L.

Everyone located between 0 and x_L will buy from Firm L. Everyone on the other side will buy from Firm R. Given that everyone buys 1 unit, the demand for Firm L's product is x_L and demand for Firm R's product is $1 - x_L$.

7.2.3 Nash Equilibrium

Given all this, what will be the price in the market? We assume that the price is determined by the Nash equilibrium. Each firm is assumed to know the strategy of the other firm. That is, each firm knows the price of their competitor. The Nash equilibrium is the price such that Firm L is unwilling to change their price given the price charged by Firm R, and Firm R is unwilling to change their price given Firm L's price.

Firm L's problem is as follows.

$$max_{p_L} \quad \frac{(v+1-\beta p_L+\beta p_R)(p_L-c_L)}{2} \tag{7.3}$$

where c_L is the marginal cost of L. Firm L's profits are equal to its demand (x_L) multiplied by its markup, price less marginal cost $(p_L - c_L)$.

The solution to the optimization problem is the solution to the first order condition.

$$\frac{v + 1 - \beta p_L + \beta p_R}{2} - \frac{\beta p_L - \beta c_L}{2} = 0 \tag{7.4}$$

Firm R's problem is similar.

$$max_{p_R} \quad \frac{(-v-1+\beta p_L-\beta p_R)(p_R-c_R)}{2} \tag{7.5}$$

The first order condition is as follows.

$$\frac{-v - 1 + \beta p_L - \beta p_R}{2} - \frac{\beta p_R - \beta c_R}{2} = 0 \tag{7.6}$$

Given these first order conditions we can write down a system of equations.

$$p_L = \frac{v+1+\beta p_R+\beta c_L}{2\beta}$$
$$\tag{7.7}$$
$$p_R = \frac{-v-1+\beta p_L+\beta c_R}{2\beta}$$

Solving the system we have the Nash equilibrium prices in the market.

$$p_L = \frac{v+1+\beta c_R+2\beta c_L}{3\beta}$$
$$\tag{7.8}$$
$$p_R = \frac{-v-1+\beta c_L+2\beta c_R}{3\beta}$$

In equilibrium, prices are determined by the relative value of the products (v), the willingness to switch between products based on price (β) and marginal costs (c_L and c_R). Note that Firm L's price is determined by Firm R's marginal costs. Do you see why that would be?

7.3   Estimating Demand in Hotelling's Model

We can illustrate the modern approach to demand estimation using Hotelling's model. The section creates a simulated the demand system based on the model and estimates the parameters using the IV approach.

7.3.1   Simulation of Hotelling Model

The simulation uses the model above to create market outcomes including prices and market shares. There are 1,000 markets. These may represent the two firms competing at different times or different places. The data is adjusted to keep market outcomes where prices are positive and shares are between 0 and 1.

```
> set.seed(123456789)
> N <- 1000 # number of markets
> beta = 0.25 # demand parameter
> v <- 3 + rnorm(N) # quality (vertical) measure
> c_L <- 3 + rnorm(N)
> c_R <- 9 + rnorm(N)
> # costs for both firms.
> p_L <- (v + 1 + beta*c_R  + 2*beta*c_L)/(3*beta)
> p_R <- (-v - 1 + beta*c_L + 2*beta*c_R)/(3*beta)
> # price function for each firm.
> x_L <- (v + 1 - beta*p_L + beta*p_R)/2
> # demand for firm L
> index <- p_L > 0 & p_R > 0 & x_L > 0 & x_L < 1
> c_L1 <- c_L[index]
> c_R1 <- c_R[index]
> p_L1 <- p_L[index]
> p_R1 <- p_R[index]
> x_L1 <- x_L[index]
> v1 <- v[index]
> # adjusting values to make things nice.
```

Figure 7.2 plots demand and relative price for product *L* from the simulated data. Do you notice anything odd about the figure? Remembering back to Econ 101, did demand slope up or down? It definitely slopes down in Equation (7.2). What went wrong?

7.3.2   Prices are Endogenous

Probably every economist in industrial organization has run a regression like what is depicted in Figure 7.2. Each one has looked at the results and has felt

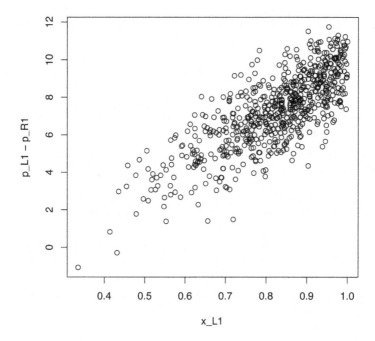

FIGURE 7.2
Demand for Firm L by relative price of Firm L to Firm R.

their heart sink because everything that they knew about economics was wrong. Then they have taken a deep breath and remembered prices are endogenous.

We are interested in estimating how prices affect demand for the product. We know they do. Equation (7.2) explicitly states that prices **cause** demand to fall. The problem is that we did not plot the demand curve. We plotted out a bunch of outcomes from the market. We plotted out a thousand equilibrium prices and equilibrium demand levels. Back to Econ 101, think of a thousand demand and supply crosses going through each of the points in Figure 7.2.

The problem in the simulation is that prices are determined endogenously. The observed market prices are the outcome of the Nash equalibria and the choices of Firm L and Firm R. If we want to estimate the demand for Firm L (Equation (7.2)) then one approach is to use the instrumental variables methods discussed in Chapter 3.

7.3.3 Cost Shifters

The standard **instrumental variable** approach is to use **cost shifters**. That is, we need an instrument related to costs. For example, if we observed c_L in the simulated data, then that would work as instrument. It satisfies the assumptions. It has a direct effect on p_L through Equation (7.8). It does not direct affect x_L, see Equation (7.2).

```
> # Intent to Treat
> lm1 <- lm(x_L1 ~ I(c_L1-c_R1))
> # First stage
> lm2 <- lm(I(p_L1-p_R1) ~ I(c_L1-c_R1))
> # IV estimate
> lm1$coefficients[2]/lm2$coefficients[2]

I(c_L1 - c_R1)
   -0.04755425
```

Remember back to Chapter 3, we can use **graph algebra** to find a simple estimator. This is the **intent to treat** regression (demand on costs) divided by the first stage regression (price on costs). Note that price of interest is the difference in the prices of the two firms. The instrument for this price is the difference in the costs of the two firms. The estimate is -0.048. The true value is -0.125, which you see from Equation (7.2) where $\beta = 0.25$ and the price effect is $\frac{-\beta}{2}$. It is not particularly accurate, but at least it is negative!

```
> y <- as.matrix(x_L1)
> X <- cbind(p_L1,p_R1)
> Z <- cbind(c_L1,c_R1)
> tab_ols <- lm_iv(y, X, Reps=500)
> tab_iv <- lm_iv(y, X, Z, Reps = 500)
> # using the function defined in chapter 3.
> row.names(tab_iv) <- row.names(tab_ols) <-
+    c("intercept",colnames(X))
> tab_res <- cbind(tab_ols[,1:2],tab_iv[,1:2])
> colnames(tab_res) <-
+    c("OLS coef","OLS sd","IV coef", "IV sd")
```

	OLS coef	OLS sd	IV coef	IV sd
intercept	0.51	0.03	1.10	0.10
p_L1	0.04	0.00	-0.04	0.01
p_R1	-0.05	0.00	0.05	0.01

TABLE 7.2
OLS and IV bootstrapped estimates of demand for the Hotelling model.

We can also use the matrix algebra method described in Chapter 3. Here we separate out the two prices and don't use the information that they have the same coefficient. We can confirm the results of Figure 7.2; relationship between price and demand have the wrong sign. The IV estimate is again on the low side at around -0.04.

7.3.4 Demand Shifters

The analysis above is the standard approach to IV in demand analysis. However, we are not limited to using **cost shifters**. In the simulation there is a third exogenous variable, v. This variable represents factors that make Firm L's product preferred to Firm R's product. We call this a **vertical** difference as opposed to the location along the line which we call **horizontal**. In the real world a vertical difference may refer to the quality of the product. Everyone prefers high quality products.

Looking at the Nash equilibrium (Equation (7.8)), we see that v positively affects the price that Firm L charges. The variable increases demand which allows the firm to charge higher prices. But this is not what we are interested in. Our goal is to estimate the slope of the demand curve. We can use the model to determine the value of interest. Equation (7.8) provides the relationship between p_L and v which is equal to $\frac{1}{3\beta}$, while the slope of demand is $-\frac{\beta}{2}$. We can use the first to determine the second. This gives -0.116, which is pretty close to -0.125. In fact, this method does better than the more standard approach.

```
> lm1 <- lm(p_L1 ~ v1)
> b_hat <- lm1$coefficients[2]
> beta_hat <- 1/(3*b_hat)
> -beta_hat/2

        v1
-0.1162436
```

The idea of using the model in this way is at the heart of much of modern industrial organization. The idea is to use standard statistical techniques to estimate parameters of the data, and then use the model to relate those data parameters to the model parameters of interest. The idea was promoted by Guerre et al. (2000) who use an auction model to back out the underlying valuations from observed bids.[1]

7.4 Berry Model of Demand

Yale industrial organization economist, Steve Berry, is probably the person most responsible for the modern approach to demand analysis. Berry et al. (1995) may be the most often used method in IO. We are not going to unpack everything in that paper. Rather, we will concentrate on the two ideas of demand inversion and taking the Nash equilibrium seriously.

We are interested in estimating demand. That is, we are interested in estimating the causal effect of price (p) on demand (or share) (s). A firm would

---

[1]The next chapter explores this approach.

like to know what would happen to its demand if it increases the price 10%. Will its revenue increase or decrease? In Chapter 5, variation in prices or other characteristics of the product are used to infer demand. The problem is that prices may not be exogenously determined.

This section presents a general IV approach to demand estimation.

7.4.1 Choosing Prices

To see how marginal costs affect price consider a simple profit maximizing firm j with constant marginal cost (c_j). Demand for the firm's product is denoted $s(p_j, p_{-j})$ (share), where p_j is their own price and p_{-j} are the prices of the competitors in the market ($-j$ means not j). The firm's pricing decision is based on maximizing its profit function.

$$\max_{p_j} \quad s_j(p_j, p_{-j})(p_j - c_j) \tag{7.9}$$

Equation (7.9) shows a firm choosing prices to maximize profits, which is quantity times margin. The solution to the maximization problem can be represented as the solution to a first order condition.

$$s_{j1}(p_j, p_{-j})(p_j - c_j) + s_j(p_j, p_{-j}) = 0 \tag{7.10}$$

where s_{j1} is the derivative with respect to p_j. This equation can be rearranged to give the familiar relationship between margins and demand.

$$\frac{p_j - c_j}{p_j} = -\frac{s_j(p_j, p_{-j})}{p_j s_{j1}(p_j, p_{-j})} = -\frac{1}{\epsilon(p_j, p_{-j})} \tag{7.11}$$

where $\epsilon(p_j, p_{-j})$ represents the elasticity of demand. For a profit maximizing firm, we can write out the optimal price margin as equal to the reciprocal of the elasticity of demand. The more elastic the demand, the lower the firm's margin.

7.4.2 A Problem with Cost Shifters

Equation (7.11) shows that as marginal cost (c) increases, price must increase. To see the exact relationship use the Implicit Function Theorem on Equation (7.10).

$$\begin{aligned} \frac{dp_j}{dc_j} &= -\frac{\frac{d.}{dc_j}}{\frac{d.}{dp_j}} \\ &= \frac{s_{j1}(p_j, p_{-j})}{s_{j11}(p_j, p_{-j})(p_j - c_j) + 2s_{j1}(p_j, p_{-j})} \end{aligned} \tag{7.12}$$

where s_{j11} is the second derivative with respect to p_j. As $s_{j1}(p_j, p_{-j}) < 0$ (demand slopes down), prices are increasing in costs. One of the IV assumptions holds, at least in theory. However, for all of the IV assumptions to hold we need linear demand or an equivalent assumption.[2] That is, we need $s_{j11}(p_j, p_{-j}) = 0$.

$$
\begin{aligned}
\frac{dp_j}{dc_j} &= \frac{s_{j1}(p_j, p_{-j})}{2s_{j1}(p_j, p_{-j})} \\
&= \frac{1}{2}
\end{aligned}
\tag{7.13}
$$

From Equation (7.12), this assumption gives a linear relationship between price and marginal cost. This is a strong assumption on the shape of demand. It is unclear if it holds in practice.

7.4.3 Empirical Model of Demand

Consider a case similar to that discussed in Chapter 5. We observe a large number of "markets" t with various products in each market. In the example below, the markets are supermarkets in particular weeks, and the products are different cereals.

For simplicity assume that there are just two products. The demand for product j in time t is the following share of the market.

$$
s_{tj} = 1 - F(\mathbf{X}'_{tj}\gamma + \beta p_{tj} + \xi_{tj})
\tag{7.14}
$$

where the share (s_{tj}) is the probability that individuals prefer product j to the alternative. The probability distribution is F. Inside F is a linear index, which looks like a standard OLS regression. The matrix $\mathbf{X}$ represents observable characteristics of the product, γ is a vector of preferences over those products, p_{tj} is the price of the product, β the preference of price and ξ_{tj} represents unobserved characteristics of the product.

This model is similar to the logit and probit models presented in Chapter 5. More formally, the model assumes that utility is **quasi-linear**. In this case, the assumption allows a neat trick. If we can invert F (and we usually can), we can write firm prices as a nice linear function of the Xs.

7.4.4 Inverting Demand

Instead of writing demand as a function of price, we can write price as a function of demand.

$$
p_{tj} = \frac{-\mathbf{X}'_{tj}\gamma}{\beta} - \frac{\xi_{tj}}{\beta} + \frac{F^{-1}(1 - s_{tj})}{\beta}
\tag{7.15}
$$

The inversion provides a nice linear relationship between price, the index over product characteristics and the inverse of the market share. Now that things are linear we can use standard IV methods. Unfortunately, things get a lot more complicated with more choices. It is not even clear that this "inversion" is always possible (Berry et al., 2013).

---

[2]See Chapter 3 for discussion of these assumptions.

In the special case of the logit demand, things are relatively straightforward (Berry, 1994).

$$s_{tj} = \frac{\exp(\mathbf{X}'_{tj}\gamma + \beta p_{tj} + \xi_{tj})}{1 + \sum_{k=1}^{J} \exp(\mathbf{X}'_{tk}\gamma + \beta p_{tk} + \xi_{tk})} \tag{7.16}$$

We can write the share of demand for product j as the logit fraction. From the logit model we can get a fairly simple linear model. Taking the inverse, that is logging both sides, we have.

$$\log(s_{tj}) = \mathbf{X}'_{tj}\gamma + \beta_j p_{tj} + \xi_{tj} - \log(1 + \sum_{k=1}^{J} \exp(\mathbf{X}'_{tk}\gamma + \beta p_{tk} + \xi_{tk})) \tag{7.17}$$

The log of share is a linear function of the utility index less information about all the other products.

Notice that it is possible to get rid of all the other characteristics. We can do this by inverting demand for the "outside" good. Remember in the logit this value is set to 1 and $\log(1) = 0$.

$$\log(s_{t0}) = \log(1) - \log(1 + \sum_{k=1}^{J} \exp(\mathbf{X}'_{tk}\gamma + \beta p_{tk} + \xi_k)) \tag{7.18}$$

From this we see the following representation. The log of the relative share is a linear function of Xs, price and the unobserved characteristics.

$$\sigma_{tj} = \log(s_{tj}) - \log(s_{t0}) = \mathbf{X}'_{tj}\gamma + \beta p_{tj} + \xi_{tj} \tag{7.19}$$

In this model, the confounding is due to the relationship between p_{tj} and ξ_{tj}.

7.4.5 Demand Shifters to Estimate Supply

If we have instruments for demand then we can rearrange the equation above.

$$p_{tj} = \frac{-\sigma_{tj} - \mathbf{X}'_{tj}\gamma - \xi_{tj}}{\beta} \tag{7.20}$$

We can write this out as an IV model presented in Chapter 3.

$$\begin{aligned} p_{tj} &= \mathbf{W}'_{tj}\delta + \xi_{tj} \\ &\text{and} \\ \mathbf{W}_{tj} &= \mathbf{Z}'_{tj}\alpha + v_{tj} \end{aligned} \tag{7.21}$$

where $\mathbf{W}$ captures transformed shares as well as **exogenous** characteristics of the product and the market and $\mathbf{Z}$ represents instruments for demand and other observables. We can now estimate demand using a standard IV estimator.

7.4.6   Demand Estimation from Supply Estimates

If all the assumptions hold, then the IV procedure above provides an estimate of the effect that changes in demand have on price. That is, the procedure estimates the slope of the supply function. But that is not what we are interested in. We want to estimate the demand function. We want to know how changes in price affect demand.

Can we use what we know about how prices are set by the firm to back out demand? Can we use game theory to back out the policy parameters of interest from the estimated parameters? Yes. This is the two-step estimation approach exemplified by Guerre et al. (2000). In the first step, we estimate a standard empirical model. In the second step, we use economic theory to back out the policy parameters of interest from the estimated parameters. Here, we estimate the slope of the supply curve and use game theory to back out the slope of the demand curve.

In order to simplify things substantially, assume that there is one product per profit maximizing firm and the demand curve is approximately linear around the optimal price $(s_{j11}(p_j, p_{-j}) = 0)$. Using the Implicit Function Theorem and Equation (7.10), the empirical relationship between prices and share can be written as a function of the slope of demand.

$$\frac{dp_j}{ds_j} = -\frac{1}{2s_{j1}(p_j, p_{-j})} \qquad (7.22)$$

The left-hand side is the observed relationship between price and demand from the data. This we can estimate with the IV procedure above. The right-hand side shows the slope of the demand function $(s_{j1}(p_j, p_{-j}))$, which is the value of interest.

This result has the following implications for the relationship between the estimated values and the parameter values of interest,

$$\hat{\beta} = -\frac{1}{2\hat{\delta}_\sigma}$$
$$\text{and} \qquad (7.23)$$
$$\hat{\gamma} = \frac{-\hat{\delta}_{-\sigma}}{\hat{\beta}}$$

where δ_σ refers to the IV estimate of the effect of changes in demand on price, and $\delta_{-\sigma}$ refers to the other exogenous parameters of the model. See Equation (7.21).

7.5   Introduction of Apple Cinnamon Cheerios

Over the last thirty years we have seen some amazing new products. The Apple iPod, the Apple iPhone, the Apple iPad, but years before any of these,

General Mills introduced Apple Cinnamon Cheerios. This product may be subject to the oddest debate in microeconometrics: what is the true value of Apple Cinnamon Cheerios? MIT econometrician, Jerry Hausman, found that the introduction of Apple Cinnamon Cheerios substantially increased consumer welfare. Stanford IO economist, Tim Bresnahan, claimed Hausman was mistaken.

I'm sure you are thinking, who cares? And you would be correct. I, myself, have never eaten Apple Cinnamon Cheerios. I am reliably informed that they are similar to Apple Jacks, but I have not eaten those either.

However, the debate did raise important issues regarding how assumptions presented above are used to estimate new products like BART or Apple Cinnamon Cheerios (Bresnahan, 1997). McFadden's approach requires that products are a sum of their attributes and preferences for those attributes is fixed across products. We will continue to use these assumptions in order to determine the value of Apple Cinnamon Cheerios.

This section uses cereal price and sales data from a Chicagoland supermarket chain in the 1980s and 1990s.

7.5.1 Dominick's Data for Cereal

Data on the demand for cereal is available from the Kilts School of Marketing at the University of Chicago. The data was collected from the Dominick's supermarket chain and stores throughout Chicagoland. We have information on 490 UPCs (products) sold in 93 stores over 367 weeks from the late 80s to the late 90s. As in Chapter 5 we want to map the products into characteristics. As there is no characteristic information other than name and product size, the Dominick's data is merged with nutritional information for 80 cereal products from James Eagan.[3] To estimate the model we need to have one product that is the "outside good." In this case, we assume that it is the product with the largest share of the products analyzed. Prices, characteristics and shares are created relative to the outside good. A more standard assumption is to classify the outside good based on a definition of the market, say "all breakfast foods." The assumption makes the exposition a lot simpler but at the cost of very strong assumptions on how individuals substitute between breakfast foods.[4]

```
> x <- read.csv("dominicks.csv", as.is = TRUE)
> p <- x$ozprice
> x$fat <- x$fat/100
> x$oz <- x$oz/100
> x$sodium <- x$sodium/1000
> x$carbo <- x$carbo/100
> # changes the scale of the variables for presentation
```

---

[3]https://perso.telecom-paristech.fr/eagan/class/igr204/datasets
[4]The cleaned data is available here: https://sites.google.com/view/microeconometrics withr/table-of-contents

```
> W <- x[,colnames(x) %in% c("sig","fat","carbo","sodium",
+                             "fiber", "oz","quaker","post",
+                             "kellogg","age9", "hhlarge")]
> # sig (sigma) refers to the adjusted measure of market share
> # discussed above
> # fat, carbo, sodium and fiber refer to cereal incredients
> # oz is the size of the package (ounces)
> # Quaker, Post and Kellogg are dummies for
> # major cereal brands
> # age9 is a measure of children in the household
> # hhlarge is a measure household size.
```

7.5.2 Instrument for Price of Cereal

Berry (1994) suggests that we need to instrument for price. Think about variation in prices in this data. Prices vary across products, as determined by the manufacturer. Prices vary across stores, as determined by the retailer (Dominick's). Prices vary across time due to sales and discounts. The last can be determined by the manufacturer or the retailer or both. The concern here is that we have variation across stores. Stores with higher demand for certain cereal products will also get higher prices.

Berry suggests that we need two types of instruments. We need instruments that exogenously vary and determine price through changes in costs. These are called **cost shifters**. They may be wages or input prices. Above it is pointed out that in theory these instruments are generally not linearly related to price. We also need instruments that vary exogenously and determine price through changes in demand. These are called **demand shifters**. They may be determined by demographic differences or by difference in product characteristics. The analysis here uses variation in income across stores. The assumptions are that Income → Demand, Demand → Price and Price = Demand(Income) + Unobservables. Remember that these are the three assumptions of an instrumental variable from Chapter 3.

```
> Z <- cbind(x$income,x$fat,x$sodium,x$fiber,x$carbo,x$oz,
+            x$age9,x$hhlarge,x$quaker,x$post,x$kellogg)
> colnames(Z) <- colnames(W)
> tab_ols <- lm_iv(p, W, Reps = 300)
> tab_iv <- lm_iv(p, W, Z, Reps = 300)
> # using the IV function from Chapter 3
> # actually using the robust version
> # presented in Appendix B.
> row.names(tab_iv) <- row.names(tab_ols) <-
+   c("intercept",colnames(W))
> tab_res <- cbind(tab_ols[,1:2],tab_iv[,1:2])
```

```
> colnames(tab_res) <-
+   c("OLS coef","OLS sd","IV coef", "IV sd")
```

	OLS coef	OLS sd	IV coef	IV sd
intercept	-0.01	0.00	0.59	2.26
sig	0.03	0.00	0.30	1.01
fat	-0.89	0.03	3.94	20.24
sodium	-0.10	0.01	-1.09	3.73
fiber	-0.01	0.00	0.02	0.10
carbo	-0.07	0.01	2.82	10.81
oz	-0.48	0.00	-0.49	0.06
age9	0.16	0.01	-0.26	1.62
hhlarge	-0.04	0.01	-0.26	1.07
quaker	-0.03	0.00	-0.17	0.52
post	-0.06	0.00	-0.03	0.09
kellogg	-0.02	0.00	-0.20	0.66

TABLE 7.3
OLS and IV bootstrapped estimates of adjusted log share on observed product and store characteristics, with income as an instrument for price in the IV model.

Table 7.3 presents the OLS and IV estimates. The OLS estimates present the non-intuitive result that price and demand are positively correlated. The IV model assumes that changes in income are **exogenous** and that they determine price *through* changes in demand. Under standard IV assumptions 0.3 measures the effect of changes in demand on price, although this is not precisely estimated. As expected, it is positive, meaning that an exogenous increase in demand is associated with higher prices. This is great, but we are interested in estimating demand, not supply.

7.5.3 Demand for Apple Cinnamon Cheerios

The discussion above suggests that we can transform the estimates from the IV model to give the parameters of interest. That is, we can use assumptions about firm behavior to back out the slope of the demand function from our estimate of the slope of the supply function. See Equation (7.23).

```
> beta_hat <- -1/(2*tab_iv[2]) # transformation into "demand"
> gamma_hat <- -tab_iv[,1]/beta_hat   # transform gammas back.
> gamma_hat[2] <- beta_hat
> # puts in the causal effect of price on demand.
> names(gamma_hat)[2] <- "price"
```

Given this transformation we can estimate the demand curve for family size Apple Cinnamon Cheerios. The following loop determines the share of each product for different relative prices of the Apple Cinnamon Cheerios.

```
> W <- as.matrix(W)
> Ts <- length(unique(x$store))
> Tw <- length(unique(x$WEEK))
> J <- length(unique(x$UPC))
> exp_delta <- matrix(NA,Ts*Tw,J)
> t <- 1
> for (ts in 1:Ts) {
+    store <- unique(x$store)[ts]
+    for (tw in 1:Tw) {
+       week <- unique(x$WEEK)[tw]
+       W_temp <- W[x$WEEK==week & x$store==store,]
+       exp_delta[t,] <- exp(cbind(1,W_temp)%*%gamma_hat)
+       t <- t + 1
+       #print(t)
+    }
+ }
> share_est <- exp_delta/(1 + rowSums(exp_delta, na.rm = TRUE))
> summary(colMeans(share_est, na.rm = TRUE))

    Min. 1st Qu.  Median    Mean 3rd Qu.    Max.
 0.01366 0.01727 0.01885 0.01873 0.02010 0.02435
```

The loop above calculates the predicted market shares for each of the products given the estimated parameters.

```
> upc_acc <- "1600062760"
> # Apple cinnamon cheerios "family size"
> share_acc <-
+    mean(x[x$UPC=="1600062760",]$share, na.rm = TRUE)
> # this is calculated to determine the relative prices.
> K <- 20
> min_k <- -6
> max_k <- -2
> # range of relative prices
> diff_k <- (max_k - min_k)/K
> acc_demand <- matrix(NA,K,2)
> min_t <- min_k
> for (k in 1:K) {
+    pr <- min_t + diff_k
+    exp_delta2 <- exp_delta
+    W_temp <- as.matrix(cbind(1, pr, W[x$UPC==upc_acc,-1]))
+    exp_delta2[x$UPC==upc_acc] <- exp(W_temp%*%gamma_hat)
+    ave_share <- matrix(NA,length(unique(x$UPC)),2)
```

```
+    for (i in 1:length(unique(x$UPC))) {
+      upc <- sort(unique(x$UPC))[i]
+      ave_share[i,1] <- upc
+      ave_share[i,2] <-
+        mean(exp_delta2[x$UPC==upc],na.rm = TRUE)
+      #print(i)
+    }
+    ave_share[,2] <-
+      ave_share[,2]/(1 + sum(ave_share[,2], na.rm = TRUE))
+    acc_demand[k,1] <- pr
+    acc_demand[k,2] <- ave_share[ave_share[,1]==upc_acc,2]
+    min_t <- min_t + diff_k
+    #print("k")
+    #print(k)
+ }
```

Figure 7.3 presents the demand curve for a family size box of Apple Cinnamon Cheerios.

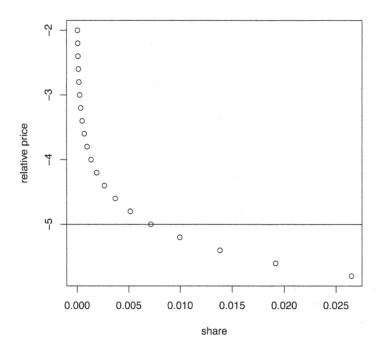

FIGURE 7.3

Plot of demand for family size Apple Cinnamon Cheerios. Line is the average relative price.

7.5.4   Value of Apple Cinnamon Cheerios

To determine the value of Apple Cinnamon Cheerios family size box we can calculate the area under the demand curve (Hausman, 1997). If we approximate the area with a triangle, we get an annual contribution to consumer welfare of around \$271,800 per year for all Dominick's customers in Chicagoland.[5] Assuming that Dominick's had a market share of 25% and that Chicagoland accounts for about $\frac{1}{32}$ of the US population, then this scales to \$35m a year. This estimate is substantially lower than Hausman's estimate of over \$60m annually for the United States. Of course, I'm unwilling to argue that my estimate is better with either Bresnahan or Hausman.

```
> (0.5*(-3-acc_demand[5,1])*acc_demand[5,2]*267888011*(52/367))
```

```
[1] 271798.6
```

7.6   Discussion and Further Reading

My field, industrial organization, is dominated by what is called **structural estimation**. That is using game theory to estimate the parameters of the model, then using the parameter estimates to make policy predictions. Industrial organization economists believe that these ideas should be used more broadly across economics.

The chapter reconsiders the problem of demand estimation. It allows that prices are determined endogenously in the model. The chapter assumes that prices are actually the result of a game played between rival firms that sell similar products. It shows that we can use a standard IV estimator to estimate the slope of the supply function, then we can use the Nash equilibrium to determine the slope of demand.

It has become standard practice in empirical industrial organization to split the estimation problem in to these two steps. The first step involves standard statistical techniques, while the second step relies on the equilibrium assumptions to back out the policy parameters of interest. Chapter 9 uses this idea to estimate the parameters of interest from auction data (Guerre et al., 2000).

Berry et al. (1995) is perhaps the most important paper in empirical industrial organization. It presents three important ideas. It presents the idea discussed above that logit demand can be inverted which allows for use of the standard instrumental variables approach. This idea is combined with an assumption that prices are determined via equilibrium of a Nash pricing game allowing the parameters of interest to be identified. If this wasn't enough, it

---

[5]The number \$267,888,011 is total revenue from the original Dominicks sales data, wcer.csv. Here we used a fraction of the original data (52/367) to get the annual amount.

adds the idea of using a flexible model called a mixed logit. All this in one paper used to estimate the demand for cars! However, without Nevo (2000) few would have understood the contributions of Berry et al. (1995). More recently Berry et al. (2013) dug into the assumptions of Berry et al. (1995) to help us understand which are important for identification and which simply simplify the model. MacKay and Miller (2019) have an excellent exposition of the various assumptions needed to estimate demand. What if firms are colluding? Fabinger and Weyl (2013) and Jaffe and Weyl (2013) are good starting places for thinking about estimating demand in that case.

8

Estimating Games

8.1 Introduction

In Chapter 7 we used game theory to revisit demand estimation. This chapter takes game theory to a broader set of problems in microeconometrics. Chapter 7 considered games with pure strategies while this chapter introduces the idea of mixed strategies. This chapter also introduces a standard estimator for game theory based models, the **generalized method of moments** (GMM). The chapter uses the GMM estimator to revisit returns to schooling and whether or not NFL coaches punt too often on 4th down.

8.2 Mixed Strategy Nash Equilibrium

In his 2006 paper, Berkeley macroeconomist David Romer argues that coaches in the NFL do not behave rationally. In particular, they choose to punt *too often* on fourth down. The goal in American football is to score points. The primary method for scoring is to run the ball across your opponent's goal-line. The ball can be moved down toward the goal-line by either running and carrying or by throwing the ball to a team-mate to catch. In American football the offense has 4 tries ("downs") to move the ball ten yards.

After the third try, the coach has a decision to make. They can "go for it." That is, try a standard play which involves running with the ball or throwing the ball. If they make the distance, the number of tries resets, which is good. If they don't make the distance, the other team gets the ball, which is bad. Most importantly, the other team gets the ball right at the same spot on the field, which is very bad. This means the other team may have a particularly easy chance to score. As I said, very bad. Punting the ball also gives the ball to the other team. But, and this is the kicker, pun intended, the other team gets the ball where the punt lands. This means that the other team may have a particularly hard chance to score, which is not that bad.

This idea that coaches punt too often has become pervasive; there is even an episode of the CBS TV show *Young Sheldon* discussing the issue. But is it correct? We do not observe the information required to evaluate the coaches' decision. The fact that coaches almost always punt means that we don't actually observe what happens when they don't punt. We don't observe the counter-factual outcome. However, we can use game theory to predict it.

8.2.1   Coaches' Decision Problem

The decision to punt or not on fourth down can be thought of as maximizing expected points rather than maximizing the probability of winning. Romer (2006) argues that, at least early on in the game, this is a reasonable way to model the coaches' problem.

Here the expected points are calculated by Ron Yurko. For more information go here: `https://github.com/ryurko/nflscrapR-data`. The data is scraped from NFL play by play information (`https://www.nfl.com/scores`).[1]

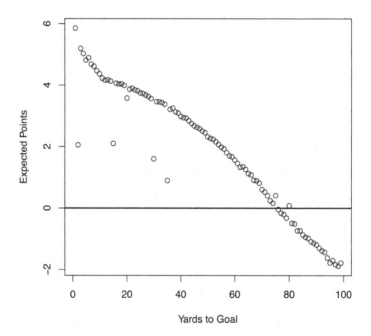

FIGURE 8.1
Plot of average expected points by yard line.

Figure 8.1 presents the average expected points for the first quarter by yard line. Expected points are calculated using **dynamic programming**. Each yard line and possession combination is a "state." We can then calculate the probability of moving from any state to any other state. For some states one

---

[1]The version of the data used here is available at `https://sites.google.com/view/microeconometricswithr/table-of-contents`.

of the teams scores. Thus we can determine the expected point differential for each yard line. Note that when a team is within their own 20, the expected points is actually negative. Even though one team may have the ball, it is more likely for the other team to get it and score.

Consider a fourth down with 5 yards to go. The ball is currently on the offense's own 45 yard line (55 yards to the goal line). There are two important numbers, the "to-go" distance of 5 yards and the position on the field which is 55 yards to the goal line. The longer the to-go distance the harder it is to get the distance. The further from the goal line, the easier it is for the other team to score if they get the ball.

Using Figure 8.1 we can calculate the expected points for the defense from each position. For illustrative purposes assume that we expect the defense to earn 0 points if the offense punts the ball. That is, the team punts the ball from their own 45 to the other team's 20. The punting earns the team -0 in expected points. Alternatively, the offense could try for the 5 yards. If they get it, they are at the 50 and they will earn 2 points in expectation. But what if they don't get it? If they turn the ball over on downs, the defense starts at the offense's 45. They have 45 yards until the goal line. This gives the offense about -2.3 in expected points. So whether to punt depends on the probability of getting the 5 yards.

The offense should go for it if the left-hand side is bigger than zero.

$$2p - 2.3(1 - p) > 0$$
$$\text{or} \tag{8.1}$$
$$p > 0.54$$

If p denotes the probability of getting the 5 yards, then the coach should go for it if $p > 0.54$. So what is p?

This seems straightforward. We just calculate the probability of getting 5 yards from the data. Just take all the fourth down attempts to go 5 yards and calculate the probabilities. The problem is that teams don't actually go for it on fourth down. We can't calculate the probability of outcomes for events that do not occur.

Romer's solution is to use the estimated probability of going 5 yards on third down. A concern with Romer's approach is third down and fourth down are different. Instead, the solution here is to use the third down data to estimate the **game** played on third down. Then take the estimated parameters and use them to predict what would occur in the game played on fourth down.

8.2.2 Zero-Sum Game

We can model the third-down game as a **zero-sum game**. These games date to the origins of game theory in the 1920s, when the mathematician John von Neumann was interested in modeling parlor games like cards or chess. In these games, there is a winner and a loser. What one person wins, another person loses.

Consider a relatively simple zero-sum game presented in Table 8.1. Remember a game has three elements; **players**, **strategies** and **payoffs**. The players are called Offense and Defense. Each has the choice between two strategies, Run and Pass. Payoffs are expected points (for simplicity). In the game, a pass play will get more yards and thus more expected points, but is generally higher risk. For the Defense, to play "Pass," means to implement a pass oriented defense such as one in which the line-backers drop back into zones. This is a **zero-sum game**, so the payoff to one player is the negative of the payoff to the other player.

{Off,Def}	Run	Pass
Run	{0,0}	{1,-1}
Pass	{2,-2}	{0,0}

TABLE 8.1
Simple zero-sum game with two players Offense and Defense, choosing between two actions, Run and Pass.

The game represented in Table 8.1 has "low" payoffs on the diagonal and "high" payoffs on the off-diagonal. That is, the Offense prefers to choose Run when the Defense chooses Pass. Similarly, they prefer Pass when the Defense chooses Run. Note also, that the payoff is higher from a successful pass than a successful run. This captures the fact that a pass tends to gain more yardage.

What is the solution to this game?

What is the **Nash equilibrium**? One possibility is that both teams play {Run, Run}. But if the Defense plays Run, it is optimal for Offense to play Pass. So that is not a Nash equilibrium. Go through the different cases. Can you find the Nash equilibrium? Is there a Nash equilibrium?

8.2.3   Nash Equilibrium

The Nobel prize winning mathematician, John Nash, proved that for relatively general games, there *always* exists at least one Nash equilibrium. However, there may not exist a **pure strategy** Nash equilibrium as was discussed above. The only Nash equilibrium may be in **mixed strategies.**

A mixed strategy is one in which the player places probability weights on the available actions. Instead of choosing Run, the player places some probability of Run and some probability of Pass. It may not be that the player literally uses a coin or some other randomization device. Rather, the assumption is that the other players in the game do not know exactly what the player will do. The other players know only the probability weights that are placed on Pass and Run.

In our problem neither team should telegraph its play choice to the other team. They should keep the other team guessing. Think about the child's game,

Rock/Paper/Scissors. You want to keep switching between the three options. It is not optimal to always choose Rock. If the other player knows you will always choose Rock, they will always choose Paper. In football, both teams try to have fake plays and formations in order to keep the other team guessing.

Determining the **mixed strategy Nash equilibrium** is tricky. You must show that the player is indifferent between her actions. In a mixed strategy Nash equilibrium, each player is indifferent between Run and Pass. If this is not true, say a player prefers Run, then mixing cannot be optimal. It is optimal to choose Run.

The Nash equilibrium in mixed strategies is where the Offense chooses a probability of Pass (q_o) such that the Defense is indifferent between Run and Pass. Similarly for q_d.

$$
\begin{aligned}
\text{Def:} &\ -2q_o = -(1-q_o) \\
\text{Off:} &\ q_d = 2(1-q_d)
\end{aligned}
\tag{8.2}
$$

The equilibrium is the solution to the two equations in Equation (8.2). The mixed strategy Nash equilibrium is $\{q_o = \frac{1}{3}, q_d = \frac{2}{3}\}$. In equilibrium, the Offense will tend to play Run, while Defense keys on Pass. Why is this? Does it make sense for the Offense to play something with a lower payoff?

8.2.4 Third and Fourth Down Game

Consider the Third Down game represented by Table 8.2. This is a **zero-sum game**, so the payoff presented is the payoff to the Offense. It is assumed that if the Offense does not make first down, then they punt. The expected points is dependent on both the current location, Y, and the expected distance of the punt ($\mathbb{E}(P|Y_k)$). This position is denoted Y_k. If the Offense is successful in getting first down, then the expected points depends on whether the play was a pass or a run, Y_p and Y_r, respectively.

{Off,Def}	Run	Pass
Run	$-(1-p_{rr})\mathbb{E}(P|Y_k)$ $+p_{rr}\mathbb{E}(P|Y_r)$	$-(1-p_{rp})\mathbb{E}(P|Y_k)$ $+p_{rp}\mathbb{E}(P|Y_r)$
Pass	$-(1-p_{rr})\mathbb{E}(P|Y_k)$ $+p_{pr}\mathbb{E}(P|Y_p)$	$-(1-p_{pr})\mathbb{E}(P|Y_k)$ $+p_{pp}\mathbb{E}(P|Y_p)$

TABLE 8.2
Zero-sum game with two players, Offense and Defense, who choose two actions, Run and Pass. The payoffs are for Offense. The payoffs for Defense are the negative.

The Third Down Game and the Fourth Down game are similar. In particular, we assume that the probabilities of success conditional on the strategies of the Offense and Defense remain the same. What changes between downs is the expected points associated with the failure. In the Third Down if the Offense fails, it gets $-\mathbb{E}(P|Y_k)$. The negative is because the ball goes to the other team. But the ball moves to the location where it finishes after the punt. In the Fourth Down game it gets $-\mathbb{E}(P|100 - Y)$. That is, the ball is given to the opponent at the exact same location. Note that they have to go the other direction.

The objective is to estimate the parameters of the game that will be used to model the policy of going for it on fourth down. Those parameters are four conditional probabilities. These are the probability of successfully getting first down conditional on the type of play, Run or Pass, chosen by both the Offense and the Defense. This set of parameters is denoted $\theta = \{p_{rr}, p_{rp}, p_{pr}, p_{pp}\}$, where p_{rp} is the probability of success when the Offense chooses Run and the Defense chooses Pass. The conditional probability, $p_{rp} = \mathrm{Pr}(\mathrm{Success}|O = \mathrm{Run}, D = \mathrm{Pass})$.

8.2.5   Equilibrium Strategies

If we could observe the actions of both the Offense and the Defense, then it would be straightforward to estimate the conditional probabilities of interest. While we have very good data on the action chosen by the Offense, we don't know what action the Defense chose. The available play by play data doesn't provide a lot of information on the type of defense that was used in each play.

To identify the conditional probabilities, we combine the observed probabilities with the constraints from the Nash equilibrium.

$$
\begin{aligned}
\mathrm{Pr}(\mathrm{Pass}) &= q_o \\
\mathrm{Pr}(\mathrm{Success}|\mathrm{Pass}) &= p_{pr}(1 - q_d) + p_{pp}q_d \\
\mathrm{Pr}(\mathrm{Success}|\mathrm{Run}) &= p_{rr}(1 - q_d) + p_{rp}q_d
\end{aligned}
\tag{8.3}
$$

In equilibrium the following equalities must hold. The probability that the Defense plays Pass (q_d) must set the expected value of Offense playing Run to the expected value of the Offense playing Pass. See the first equality. Similarly, the probability that the Offense plays Pass (q_o) must set the expected value of Defense playing Run to the expected value of the Defense playing Pass.

$$
\begin{aligned}
(1 - q_d)V_{3rr} + q_d V_{3rp} &= (1 - q_d)V_{3pr} + q_d V_{3pp} \\
(1 - q_o)V_{3rr} + q_o V_{3pr} &= (1 - q_o)V_{3rp} + q_o V_{3pp}
\end{aligned}
\tag{8.4}
$$

where the value functions are as follows.

$$
\begin{aligned}
V_{3rr} &= -(1 - p_{rr})\mathbb{E}(P|Y_k) + p_{rr}\mathbb{E}(P|Y_r) \\
V_{3rp} &= -(1 - p_{rp})\mathbb{E}(P|Y_k) + p_{rp}\mathbb{E}(P|Y_r) \\
V_{3pr} &= -(1 - p_{rp})\mathbb{E}(P|Y_k) + p_{pr}\mathbb{E}(P|Y_p) \\
V_{3pp} &= -(1 - p_{pp})\mathbb{E}(P|Y_k) + p_{pp}\mathbb{E}(P|Y_p)
\end{aligned}
\tag{8.5}
$$

Assuming these equalities hold in the data allows us to estimate the parameters.

Rearranging we have the equilibrium strategies for the Offense and Defense.

$$q_o = \frac{V_{3rp} - V_{3rr}}{(V_{3rp} - V_{3rr}) + (V_{3pr} - V_{3pp})}$$

$$q_d = \frac{V_{3pr} - V_{3rr}}{(V_{3pr} - V_{3rr}) + (V_{3rp} - V_{3pp})}$$

(8.6)

In general, the expected points are larger when there is a mis-match in the actions of the two teams. This suggests that the strategies will always lie between 0 and 1. Can you provide an intuitive explanation of these probabilities? It is not at all obvious. For example if the value to the Offense of playing Pass versus Run increases then the probability the Offense plays Pass *decreases*.

8.2.6 Equilibrium Strategies in R

The estimator involves solving for the equilibrium strategies of the Offense and Defense. The function takes in the parameters (the conditional probabilities) and the expected points for each situation. It then calculates the equilibrium probability of playing Pass for the Offense and the Defense based on Equation (8.6). Note that the expected points associated with failure depend on which down. For third down it is assumed that it is determined by the expected points from a punt, while for fourth down it is the expected points from turning the ball over to the other team at that position.

```
> # a function for determining the equalibrium strategies.
> q_fun <- function(theta,Ep) {
+    Ep_run <- Ep[,1]
+    Ep_pass <- Ep[,2]
+    Ep_fail <- Ep[,3]
+    V_rr <- - (1 - theta[1])*Ep_fail + theta[1]*Ep_run
+    V_rp <- - (1 - theta[2])*Ep_fail + theta[2]*Ep_run
+    V_pr <- - (1 - theta[3])*Ep_fail + theta[3]*Ep_pass
+    V_pp <- - (1 - theta[4])*Ep_fail + theta[4]*Ep_pass
+    q_o <- (V_rp - V_rr)/((V_rp - V_rr) + (V_pr - V_pp))
+    q_d <- (V_pr - V_rr)/((V_pr - V_rr) + (V_rp - V_pp))
+    for (i in 1:length(q_o)) { q_o[i] <- min(max(q_o[i],0),1) }
+    for (i in 1:length(q_d)) { q_d[i] <- min(max(q_d[i],0),1) }
+    # forcing results to be probabilities.
+    return(list(q_o=q_o,q_d=q_d))
+ }
```

8.2.7 Simulation of Third and Fourth Down Game

To illustrate the estimator, consider a simulation of the Third Down game presented above. We observe a similar game played 2,000 times.

```
> set.seed(123456789)
> N <- 2000 # number of plays
> theta <- c(0.2, 0.8, 0.5, 0.1)
> # parameters of the model
> # conditional probabilities
> Y <- 50 + 20*runif(N)
> # current yards to go.
> Y_k <- Y - rnorm(N,mean=35)
> Y_k <- ifelse(Y_k > 0, Y_k, 20)
> # yards to go after punt.
> Y_p <- Y - rnorm(N,mean=15)
> Y_p <- ifelse(Y_p > 0, Y_p,0)
> # yards to go after pass
> Y_r <- Y - rnorm(N,mean=3)
> Y_r <- ifelse(Y_r > 0, Y_r,0)
> # yards to go after run
> EP <- function(x) 5 - 5*x/100
> # an expected points function to approximate figure above.
> # equalibrium strategies
> q_3 = q_fun(theta, cbind(EP(Y_r), EP(Y_p), EP(100 - Y_k)))
> q_4 = q_fun(theta, cbind(EP(Y_r), EP(Y_p), EP(100 - Y)))
```

	3rd O Pass	3rd D Pass	4th O Pass	4th D Pass
1	Min. :0.5444	Min. :0.3399	Min. :0.5629	Min. :0.3269
2	1st Qu.:0.5572	1st Qu.:0.3577	1st Qu.:0.5711	1st Qu.:0.3390
3	Median :0.5604	Median :0.3627	Median :0.5733	Median :0.3423
4	Mean :0.5604	Mean :0.3627	Mean :0.5733	Mean :0.3423
5	3rd Qu.:0.5636	3rd Qu.:0.3678	3rd Qu.:0.5754	3rd Qu.:0.3457
6	Max. :0.5748	Max. :0.3881	Max. :0.5830	Max. :0.3588

TABLE 8.3
Summary stats of the equilibrium strategies for the simulated data by down.

Note the difference between third down and fourth down. Table 8.3 shows that the change in payoffs leads to a subtle change in the strategies for the simulated teams. For the Offense there is a 1 percentage point increase in the probability of Pass, while for the Defense there is a 2 to 3 percentage point *decrease* in the probability of Pass. Do you see why that would be? Why does the Defense key on Run when moving to 4th down?

```
> Pass <- runif(N) < q_3$q_o
> success3 <-
+   runif(N) <
+   Pass*(theta[3]*(1 - q_3$q_d) + theta[4]*q_3$q_d) +
```

```
+    (!Pass)*(theta[1]*(1 - q_3$q_d) + theta[2]*q_3$q_d)
> play = ifelse(Pass,"pass","run")
> # this is created to simulate the estimator below.
> # fourth down
> Pass <- runif(N) < q_4$q_o
> success4 <- runif(N) <
+    Pass*(theta[3]*(1 - q_4$q_d) + theta[4]*q_4$q_d) +
+    (!Pass)*(theta[1]*(1 - q_4$q_d) + theta[2]*q_4$q_d)
```

The Defense moves to become more focused on stopping the Run. This change in the strategies leads to a change in the success rates. In particular, the probability of the Offense successfully completing a pass or run falls by more than a percentage point. These simulated results suggest that using third down success rates leads to biased estimates and makes "going for it" look more valuable than it actually is.

```
> mean(success3)
```

```
[1] 0.3865
```

```
> mean(success4)
```

```
[1] 0.369
```

8.3   Generalized Method of Moments

This section introduces a new estimation algorithm, **generalized method of moments** (GMM). The algorithm was developed by Nobel prize winning economist, Lars Peter Hansen. While its initial applications were in macroeconomics, GMM has become standard in microeconometric estimation problems involving game theory. In order to understand the algorithm, we will take a detour and return to the question of estimating OLS.

8.3.1   Moments of OLS

The algorithm is a generalization of the least squares algorithm presented in Chapter 1. OLS can be estimated by noting that the mean of the error term is equal to zero. That is, when the first **moment** of the error term distribution is zero.

In the model introduced in Chapter 1, we have some outcome that is a function of observables and unobservables.

$$y_i = \mathbf{X}_i'\beta + v_i \tag{8.7}$$

where y_i is the observed outcome of interest for unit i, $\mathbf{X}_i$ are the vector of observed **explanatory variables**, β is the vector of parameters of interest and v_i represents unobserved characteristics of unit i.

Let the unobserved term be normally distributed, $v_i \sim \mathcal{N}(0, \sigma^2)$. The first moment of the unobserved characteristic is equal to 0.

$$\mathbb{E}(\epsilon_i) = \mathbb{E}(y_i - X_i'\beta) = 0 \tag{8.8}$$

The second moment is the square of the unobserved term which is equal to variance (σ^2).

$$\mathbb{E}(\epsilon_i^2) - \sigma^2 = \mathbb{E}((y_i - X_i'\beta)^2) - \sigma^2 = 0 \tag{8.9}$$

Note that in this case $(\mathbb{E}(\epsilon_i))^2 = 0.$[2]

In Chapter 1 we saw that we can use least squares to estimate β using the first moment. This suggests that we can also use least squares to estimate the variance σ^2.

$$\min_{\hat{\beta},\hat{\sigma}} \quad \sum_{i=1}^{N}((y_i - \mathbf{X}_i'\hat{\beta})^2 - \hat{\sigma}^2)^2 \tag{8.10}$$

Hansen noted that in various problems there may be more than one **moment** that must equal to zero. Having multiple moment conditions seems great. But it may be too much of a good thing. Multiple moments imply multiple solutions. While in theory there may exist only one set of parameters that satisfy all the moment conditions, in data there may exist many different parameters that satisfy the conditions. Below, we discuss Hansen's solution to this problem.

8.3.2   Simulated Moments OLS

Consider the data we simulated in Chapter 1. This time there is a slight difference; in this data $\sigma = 2$ rather than 1. The two functions below are based on the equations presented above. The first estimates OLS using the first moment of the unobserved distribution and the second estimates OLS using the second moment. Using the second moment allows all three parameters to be estimated.

```
> set.seed(123456789)
> N <- 500
> a <- 2
> b <- 3
> x <- runif(N)
> u <- rnorm(N, mean=0, sd=2)
> y <- a + b*x + u
> f_1mom <- function(beta, y, X) {
+    y <- as.matrix(y)
+    X <- as.matrix(cbind(1,X))
```

---

[2]Variance of a random variable x is $\mathbb{E}((x - \mu)^2)$, where $\mathbb{E}(x) = \mu$.

```
+    sos <- mean((y - X%*%beta)^2)
+    return(sos)
+ }
> f_2mom <- function(par, y, X) {
+    y <- as.matrix(y)
+    X <- as.matrix(cbind(1,X))
+    sigma <- exp(par[1])  # use to keep positive
+    beta <- par[-1]
+    sos <- mean(((y - X%*%beta)^2 - sigma^2))^2
+    return(sos)
+ }
> a <- optim(par=c(2,3), fn = f_1mom, y=y, X=x)
> # beta
> a$par

[1] 1.993527 3.027305

> b <- optim(par=c(log(2),2,3), fn = f_2mom, y=y, X=x)
> # sigma
> exp(b$par[1])

[1] 2.032743

> # beta
> b$par[-1]

[1] 1.556888 3.385584
```

We can estimate OLS using two different moment conditions, but that also gives us two different answers, although neither answer is particularly accurate. The first moment estimator gives estimates of the intercept and slope that are pretty close to the true values of 2 and 3. The second moment estimator also estimates the variance. The estimate is close to the true value of 2, but it does a poor job estimating the intercept and slope.

Can we improve on these estimates by combining the two moment estimators? The most obvious way to do this is to add them together.

```
> f_gmm_simple <- function(par, y, X) {
+    y <- as.matrix(y)
+    X <- as.matrix(cbind(1,X))
+    sigma <- exp(par[1])
+    beta <- par[-1]
+    sos <-  mean((y - X%*%beta)^2) +
+                  mean(((y - X%*%beta)^2 - sigma^2)^2)
+    return(sos)
+ }
> c <- optim(par=c(log(2),2,3), fn = f_gmm_simple, y=y, X=x)
```

```
> # sigma
> exp(c$par[1])
```

[1] 2.014251

```
> # beta
> c$par[-1]
```

[1] 1.902530 3.151387

That is, we could equally weight the two conditions. This gives an estimate of the variance and estimates of the intercept and slope that average over the previous two results. The variance estimate is pretty good but the intercept and slope estimates are not particularly close to the true values.

Why use equal weights? Why not use some other weights? Which weights should we use?

8.3.3   GMM OLS Estimator

Let $\theta = \{\beta, \sigma\}$ represent the parameters we are trying to estimate. Let each moment condition be denoted $g_{ki}(\theta, y_i, \mathbf{X}_i)$, and $g_i(\theta, y_i, \mathbf{X}_i)$ denote the vector of moment conditions. So by definition we have $\mathbb{E}(g_{ki}(\theta, y_i, \mathbf{X}_i)) = 0$ for all $k \in \{1, 2\}$ and $i \in \{1, ...N\}$.

The analog estimator is then one that finds smallest values for the vector of moment conditions. The estimator minimizes the following analog:

$$\hat{\theta} = \arg\min_{\theta} \left(\frac{1}{N} \sum_{i=1}^{N} g_i(\theta, y_i, \mathbf{X}_i) \right)' \mathbf{W} \left(\frac{1}{N} \sum_{i=1}^{N} g_i(\theta, y_i, \mathbf{X}_i) \right) \qquad (8.11)$$

where $\mathbf{W}$ is a 2×2 positive semi-definite matrix. This matrix provides the weights. Hansen (1982) shows that an optimal weighting matrix is a function of the true parameter values, which are unknown. The estimate of the weighting function is then the appropriate sample analog.

$$\hat{\mathbf{W}} = \left(\frac{1}{N} \sum_{i=1}^{N} g_i(\hat{\theta}, y_i, \mathbf{X}_i) g_i(\hat{\theta}, y_i, \mathbf{X}_i)' \right)^{-1} \qquad (8.12)$$

Below the estimation procedure will determine $\hat{\theta}$ and $\hat{\mathbf{W}}$ simultaneously.

Note that the notation here is pretty confusing. In particular, it is hard to keep track of the different summations and what exactly is going on with the vectors. Part of the confusion is that the ordering is different between Equation (8.11) and Equation (8.12). In Equation (8.11) we take the mean first so we have a vector of means; then we multiply those together. In Equation (8.12) we multiply vectors together at the observation level and take the mean of those.

It is easier to see the difference to be re-writing Equation (8.12) using matrices.

$$\hat{\mathbf{W}} = \frac{1}{N}(\mathbf{GG'})^{-1} \qquad (8.13)$$

where $\mathbf{G}$ is a $K \times N$ matrix where each column is the vector $g_i(\theta, y_i, \mathbf{X}_i)$.

8.3.4 GMM OLS Estimator in R

The GMM estimator is in two parts. There is a general GMM function that takes in the $\mathbf{G}$ matrix. This is a matrix where each row is particular moment. Note that the function is written with a check on the number of moments. The second part is the particular GMM example. In this case it is the GMM estimator for OLS using the first and second moments of the unobserved characteristic distribution.

```
> f_gmm <- function(G, K) {
+    G <- as.matrix(G)
+    N <- dim(G)[2]
+    if (K==dim(G)[1]) {
+       # a check that the matrix G has K rows
+       g <- rowMeans(G, na.rm = TRUE)
+       W <- try(solve(G%*%t(G)/N), silent = TRUE)
+       # try() lets the function work even if there is an error
+       if (is.matrix(W)) {
+          # if there is no error, W is a matrix.
+          return(t(g)%*%W%*%g)
+       }
+       else {
+          # allow estimation assuming W is identity matrix
+          return(t(g)%*%g)
+       }
+    }
+    else {
+       return("ERROR: incorrect dimension")
+    }
+ }
> f_ols_gmm <- function(par, y, X) {
+    y <- as.matrix(y)
+    X <- as.matrix(cbind(1,X))
+    sigma <- exp(par[1])
+    beta <- par[2:length(par)]
+    g1 <- y - X%*%beta
+    g2 <- (y - X%*%beta)^2 - sigma^2
+    return(f_gmm(t(cbind(g1,g2)),K=2))
+ }
```

```
> d <- optim(par=c(log(2),2,3), fn = f_ols_gmm, y=y, X=x)
> exp(d$par[1])
```

[1] 2.013754

```
> d$par[2:3]
```

[1] 1.969362 3.075517

The GMM estimator does a pretty good job. It estimates the variance parameter relatively well and does a better job at estimating the intercept and slope. It is not quite as good as least squares for β, but it is the only estimator that is able to estimate both β and σ with reasonable accuracy.

8.3.5   GMM of Returns to Schooling

A standard use of GMM is as an instrumental variable estimator. In particular, GMM can be used when we have multiple instruments for the same variable. We saw this in Chapter 3. We have two potential instruments for the level of education, distance to college and parents at home. In Chapter 3 we used these to conduct an over-identification test. Which they passed! More accurately, which they didn't fail!

Above we used the first and second moment of the distribution of the unobserved term to create our GMM estimator. Here, we use a moment of the joint distribution between the instrument and the unobserved term. Recall an important assumption of an instrument. It is independent of the unobserved term. In the graph, there is no arrow from the unobserved term to the instrument.

One implication of this assumption is that the unobserved term and the instrument are *not* correlated.

$$\mathbb{E}(z_i v_i) = 0 \tag{8.14}$$

where z_i is a proposed instrument. Note that we can re-write this in the following way.

$$\mathbb{E}(z_i(y_i - \mathbf{X}_i'\beta)) = 0 \tag{8.15}$$

Further, if we replace z_i with one of the explanatory variables x_i then we have an alternative OLS estimator.

To see how it works we can return to returns to schooling and the data we used in the first three chapters from Card (1995). The code is identical to the code in Chapter 3. The difference is that instead of using one instrument for level of education, we can use two. Note that for simplicity I don't instrument for experience.

```
> y <- x1$lwage76
> X <- cbind(x1$ed76, x1$exp, x1$exp2, x1$black, x1$reg76r,
```

```
+                x1$smsa76r, x1$smsa66r, x1$reg662, x1$reg663,
+                x1$reg664, x1$reg665,x1$reg666, x1$reg667,
+                x1$reg668, x1$reg669)
> x1$age2 <- x1$age76^2

> f_iv_gmm <- function(beta, y, X, Z) {
+    y <- as.matrix(y)
+    X <- as.matrix(cbind(1,X))
+    Z <- as.matrix(Z) # matrix of instruments of schooling
+    g1 <- Z[,1]*(y - X%*%beta)
+    g2 <- Z[,2]*(y - X%*%beta)
+    return(f_gmm(t(cbind(g1,g2)),K=2))
+ }
```

As a reminder, we can compare the IV estimator to the OLS estimator presented in Chapters 2 and 3. The estimated parameters from the OLS estimator are used as starting values for the GMM estimator.

```
> X1 <- cbind(1,X)
> beta_hat <- solve(t(X1)%*%X1)%*%t(X1)%*%y
> beta_hat[2]

[1] 0.07469326

> a <- optim(par=beta_hat, fn = f_iv_gmm, y=y, X=X,
+            Z=cbind(x1$nearc4,x1$momdad14))
> a$par[2]

[1] 0.07350604
```

The GMM estimator allows both distance to college and parents at home to instrument for education. Interestingly, this estimator gives *lower* values for returns to schooling than OLS. This is in contrast to the IV results presented in Chapter 3 and the Heckman estimates presented in Chapter 6.

8.4   Estimating the Third Down Game

After a detour to learn more about GMM, we can now use the estimator for estimating our third down game. Remember, we are using the mixed strategy Nash equilibrium to generate moments that we can use to estimate the parameters.

8.4.1   Moment Conditions

Equations (8.3) and (8.4) suggest a method for estimating the parameters of interest. The mixed strategy Nash equilibrium provides the Offense and Defense strategies conditional on the expected points for each option and the conditional probabilities.

$$
\begin{aligned}
\mathbb{E}(s_i|O = \text{Pass}) &= p_{pr}(1 - q_d(\theta, V(\mathbf{X}_i))) + p_{pp}q_d(\theta, V(\mathbf{X}_i))) \\
\mathbb{E}(s_i|O = \text{Run}) &= p_{rr}(1 - q_d(\theta, V(\mathbf{X}_i))) + p_{rp}q_d(\theta, V(\mathbf{X}_i)))) \\
\text{Pr}(O = \text{Pass}) &= q_o(\theta, V(\mathbf{X}_i))
\end{aligned} \tag{8.16}
$$

where $s_i \in \{0, 1\}$ is indicator of success, $\theta = \{p_{rr}, p_{rp}, p_{pr}, p_{pp}\}$ is the vector representing the parameters of interest and $\mathbf{X}_i$ represents observable characteristics of the situation. The functions $q_o(.)$ and $q_d(.)$ represent the equilibrium strategies of the Offense and Defense given the conditional probabilities and expected points.

The first condition states that conditional on the Offense playing Pass, the predicted success rate must be the same as the observed rate in expectation. The second condition is similar but for when the Offense plays Run. The third condition states that the observed probability of the Offense playing pass on third down must be equal to the predicted probability from the third down game, on average.

Equation (8.17) presents the sample analogs of the moment conditions in Equation (8.16).

$$
0 = \tfrac{1}{N} \sum_{i=1}^{N}(\mathbb{1}(\text{Pass}_i = 1)(s_i - \theta_3(1 - q_d(\theta)) - \theta_4 q_d(\theta)))
$$

$$
0 = \tfrac{1}{N} \sum_{i=1}^{N}(\mathbb{1}(\text{Pass}_i = 0)(s_i - \theta_1(1 - q_d(\theta)) - \theta_2 q_d(\theta))) \tag{8.17}
$$

$$
0 = q_o(\theta) - \tfrac{1}{N} \sum_{i=1}^{N} \mathbb{1}(\text{Pass}_i = 1)
$$

The GMM estimator finds the vector $\hat{\theta}$ that minimizes the weighted average of these three moments.

8.4.2   Third Down GMM Estimator in R

Assumption 6. *For all* $|Y - Y'| < \epsilon$,

$$
\text{Pr}(Success|O, D, Y, \mathbf{X}) = \text{Pr}(Success|O, D, Y', \mathbf{X}) \tag{8.18}
$$

Assumption 6 allows the parameters to be estimated with variation in the situation, but holding the parameters constant. It states that for small changes in the yardage, the success probabilities are unchanged conditional on the yardage, the actions of the Offense and Defense, and the observed characteristics, $\mathbf{X}$.

The GMM estimator has three parts. The first part of the estimator assumes that Equation (8.4) holds. It uses the Nash equilibrium to determine mixed strategies of the Offense and Defense given the parameters and the expected outcomes from each of the choices. The second part is the analog of the moment condition. The last part determines the estimated weighting matrix conditional on the observed probabilities, the expected points and the estimated parameter values.

```
> # the GMM estimator, which calls the general GMM
> # function above.
> p3_fun <- function(par,Ep,s,play) {
+    theta <- exp(par)/(1 + exp(par))
+    # using sigmoid function to keep values between 0 and 1
+    q_3 <- q_fun(theta,Ep)
+    # determine the equalibrium strategies.
+    # moments
+    Pass <- play=="pass"
+    g1 <- Pass*(s - theta[3]*(1 - q_3$q_d) - theta[4]*q_3$q_d)
+    g2 <-
+       (!Pass)*(s - theta[1]*(1 - q_3$q_d) - theta[2]*q_3$q_d)
+    g3 <- Pass - q_3$q_o
+    G <- t(cbind(g1,g2,g3))
+    # note the transpose.
+    return(f_gmm(G,3))
+ }

> EP1 <- cbind(EP(Y_r), EP(Y_p), EP(100 - Y_k))
> a <- optim(par=log(2*theta),fn=p3_fun,Ep=EP1,
+               s=success3,play=play,control = list(maxit=10000))
> exp(a$par)/(1 + exp(a$par))

[1] 0.2740818 0.6630758 0.4570891 0.2010548
```

The estimator does an okay job. The true values are $\theta = \{0.2, 0.8, 0.5, 0.1\}$. None of the parameters are estimated particularly accurately. What happens if you use a larger data set?

8.5 Are NFL Coaches Rational?

The assumption of decision maker rationality underpins many of the models in macroeconomics and microeconomics, including estimators presented in this book. Romer (2006) argues that NFL coaches are not behaving rationally, and that this has implications for the foundations of economics.

The difficulty with testing the rationality of these decisions is that NFL coaches do not actually go for it on fourth down. Therefore, we cannot actually measure what happens. We use game theory to model third and fourth downs. In the model, the success rates depend upon success rates conditional on the strategies of the Offense and Defense. We use the observed third down information to estimate these conditional success rates. We then use these estimates and the model to determine what the success rates would have been if the coach had decided to go for it on fourth down.

We can determine the "rationality" of NFL coaches by comparing the predicted rate of going for it against the actual rate. If coaches are rational, then the predicted rate of going for it should not be terribly different from the actual rate of going for it. Of course, this is based on the enormous assumption that the econometrician knows more about NFL than an NFL coach! Or at least, the model used here is a reasonable facsimile of third down and fourth down situations.

8.5.1   NFL Data

In order to estimate the parameters of the game we need to estimate the success rates, the Offense's strategy, and the value functions. The next play listed in the data is assumed to be the next play that occurs in the game. We will use the next play to determine the "result" of the play.

```
> x <- read.csv("NFLQ1.csv", as.is = TRUE)
> x$id <- as.numeric(row.names(x))
> x$res_pos <- c(x$posteam[2:dim(x)[1]],NA)
> x$res_ep <- c(x$ep[2:dim(x)[1]],NA)
> x$res_ep <- ifelse(x$posteam==x$res_pos,x$res_ep,-x$res_ep)
> x$res_game <- c(x$game_id[2:dim(x)[1]],NA)
> x$res_down <- c(x$down[2:dim(x)[1]],NA)
> x$diff_ep <- x$res_ep - x$ep
> x$diff_ep <- ifelse(x$game_id==x$res_game,x$diff_ep,NA)
> x$year <- sub("-.*","",x$game_date)
> # this uses "real expressions."
> # it subs out everything after the "-".
> x$succ <- NA
> x$succ <- ifelse(x$res_down==1 & x$posteam==x$res_pos &
+                  x$down==3,1,x$succ)
> x$succ <- ifelse(x$res_down==4,0,x$succ)
> x$pct_field <- x$yardline_100/100
> x$year <- as.numeric(x$year)
```

8.5.2 Estimating Third Down Game in R

```
> lm_run <- lm(diff_ep ~ ydstogo + pct_field + year,
+                 data=x[x$play_type=="run" &
+                     x$down==3 & x$succ==1,])
> lm_pass <- lm(diff_ep ~ ydstogo + pct_field + year,
+                  data=x[x$play_type=="pass" &
+                      x$down==3 & x$succ==1,])
> lm_punt <- lm(diff_ep ~ ydstogo + pct_field + year,
+              data=x[x$play_type=="punt",])
```

The next step is to determine the expected points after each action that the Offense could take; pass, run or punt.[3] The first regression is on the change in expected points after a third down run that gets a first down. Table 8.4 presents the results from the three OLS regressions on difference in expected points. The results show that successful pass plays get much larger increases in expected points than successful run plays. As expected these effects are larger the further to go to first down and the further to go to goal.

Given these results we can determine the predicted expected points for a run, a pass, and a punt. Note that these are calculated for every play in the data. Later, we will only use this information on the relevant subset.

```
> X <- as.matrix(cbind(1,x$ydstogo,x$pct_field,x$year))
> x$pass_ep <- x$ep + X%*%lm_pass$coefficients
> x$run_ep <- x$ep + X%*%lm_run$coefficients
> x$punt_ep <- x$ep + X%*%lm_punt$coefficients
```

We can use GMM to estimate the parameters from the Third Down game. This is done for "to go" distances of 1 to 4 yards and position from the team's own 30 yard line to the opponent's 40 yard line (30 yards). Given the sparsity of data we include data from ten yards either side of the position of interest and 1 yard either side of the to-go distance.[4]

```
> theta_hat <- matrix(NA,4*30,4)
> k <- 1
> for (i in 1:4) {
+    tg <- i # to go distance.
+    for (j in 1:30) {
+       yl <- 29 + j # yardline
+       # create subset for analysis
+       index <- x$down==3 &
+         x$yardline_100 > yl & x$yardline_100 < (yl+20) &
+         x$ydstogo > tg - 2 & x$ydstogo < tg + 2 &
+         (x$play_type=="pass" | x$play_type=="run")
```

---

[3]For simplicity we won't consider options to kick a field goal.
[4]You should try different values.

		Dependent variable:	
		diff_ep	
	(1)	(2)	(3)
ydstogo	0.220***	0.153***	0.019***
	(0.005)	(0.003)	(0.003)
pct_field	0.493***	1.808***	1.874***
	(0.069)	(0.053)	(0.102)
year	0.004	−0.005	0.031***
	(0.005)	(0.004)	(0.005)
Constant	−8.146	10.331	−63.722***
	(10.095)	(7.675)	(10.314)
Observations	1,411	3,896	5,860
R^2	0.604	0.503	0.076
Note:		*p<0.1; **p<0.05; ***p<0.01	

TABLE 8.4
OLS regression results on change in expected points due to a successful run
on third down (1), a successful pass on third down (2) and a punt on fourth
down (3).

```
+       y <- x[index,]
+       index_na <-
+         is.na(rowSums(cbind(y$run_ep,y$pass_ep,y$punt_ep,
+                       y$succ,
+                       (y$play_type=="pass" |
+                           y$play_type=="run"))))==0
+       # GMM to determine parameters (conditional probabilities)
+       y_na <- y[index_na,]
+       a <- optim(par=rnorm(4,mean=0,sd=2),fn=p3_fun,
+                   Ep=cbind(y_na$run_ep,y_na$pass_ep,
+                       y_na$punt_ep),
+               s=y_na$succ,play=y_na$play_type,
+                   control = list(maxit=10000))
+       theta_hat[k,] <- exp(a$par)/(1 + exp(a$par))
+       k <- k + 1
+       #print(k)
+   }
+ }
```

8.5.3 Predicting the Fourth Down Game

We can use the parameter estimates from the third down game to model the fourth down game. The purpose is to model the policy of going for it at various positions on the field and to go distances. The model assumes that both Offense and Defense play a mixed strategy Nash equilibrium.

The game is almost identical to the third down game (Table 8.2); the difference is the payoffs from the unsuccessful play. In the third down game, the payoff is the expected points from punting the ball away. Here it is the expected points from turning the ball over on downs. That is, the expected points if the ball goes to the other team and that team gets to start where the ball is currently located. The question is whether this one change to the payoffs substantially changes the Nash equilibrium and the predicted outcome of the game. Note that other expected point calculations and the conditional probabilities are all the same as for third down, with the conditional probabilities determined from the GMM procedure.

```
> lm_over <- lm(ep ~ pct_field + I(pct_field^2),
+                data=x[x$down==1,])
> x$over_ep <- -cbind(1,(1-x$pct_field),
+                (1-x$pct_field)^2)%*%lm_over$coefficients
```

To estimate the expected points from a turnover on downs, we use the expected points at first down from various distances to goal. The OLS regression assumes that there is a non-linear relationship.

```
> p4_fun <- function(theta,Ep) {
+    q_4 <- q_fun(theta,Ep)
+    p_4p <- q_4$q_o*(theta[3]*(1 - q_4$q_d) + theta[4]*q_4$q_d)
+    p_4r <-
+       (1 - q_4$q_o)*(theta[1]*(1 - q_4$q_d) + theta[2]*q_4$q_d)
+    p_4 <- p_4r + p_4p
+    Epgfi <- p_4r*Ep[,1] + p_4p*Ep[,2] + (1 - p_4)*Ep[,3]
+    # expected points going for it.
+    return(list(p_4=p_4,Epgfi=Epgfi))
+ }

> tab_res <- prob_actual <- prob_pred <- matrix(NA,4,30)
> k <- 1
> for (i in 1:4) {
+    tg <- i # to go distance.
+    for (j in 1:30) {
+       yl <- 29 + j # yardline
+       # create subset for analysis
+       index3 <- x$down==3 &
+          x$yardline_100 > yl & x$yardline_100 < (yl+20) &
+          x$ydstogo > tg - 2 & x$ydstogo < tg + 2 &
```

```
+        (x$play_type=="pass" | x$play_type=="run")
+        y <- x[index3,]
+        # determine predicted success on 4th down.
+        succ4 <- p4_fun(theta_hat[k,],
+                        cbind(y$run_ep,y$pass_ep,y$over_ep))
+        # Actual frequency of going for it on 4th down.
+        index4 <- x$down==4 &
+          x$yardline_100 > yl & x$yardline_100 < (yl+20) &
+          x$ydstogo > tg - 2 & x$ydstogo < tg + 2
+        z <- x[index4,]
+        z$go <- ifelse(z$play_type=="run" |
+                       z$play_type=="pass",1,NA)
+        z$go <- ifelse(z$play_type=="punt",0,z$go)
+        # relative value of punting
+        tab_res[i,j] <- mean(y$punt_ep - succ4$Epgfi,
+                             na.rm = TRUE)
+        # predicted probability of going for it
+        prob_pred[i,j] <- mean(y$punt_ep - succ4$Epgfi < 0,
+                               na.rm = TRUE)
+        # actual probability of going for it.
+        prob_actual[i,j] <- mean(z$go,na.rm = TRUE)
+        k <- k + 1
+        #print(k)
+    }
+ }
```

We can estimate the model at different distances to goal and yards to go
to determine whether it is better to punt or "go for it" on fourth down.

	1 To Go	2 To Go	3 To Go	4 To Go
1	Min. :0.2307	Min. :0.1974	Min. :-0.1795	Min. :-0.1857
2	1st Qu.:0.7204	1st Qu.:0.5967	1st Qu.: 1.0561	1st Qu.: 1.5637
3	Median :0.9009	Median :0.9985	Median : 1.2963	Median : 1.6158
4	Mean :1.4960	Mean :1.0110	Mean : 1.4380	Mean : 1.8317
5	3rd Qu.:1.8238	3rd Qu.:1.3690	3rd Qu.: 1.5271	3rd Qu.: 1.7396
6	Max. :5.3896	Max. :2.8439	Max. : 3.3797	Max. : 3.9054

TABLE 8.5
Summary stats of the expected points for punting less the expected points for
going for it for 1 yard to go up to 4 yards to go.

The table suggests that is almost always better to punt the ball away on
fourth down. Table 8.5 presents summary statistics on the difference between
expected points from punting over going for it, at each to go distance from 1
yard to 4 yards.

8.5.4 Testing Rationality of NFL Coaches

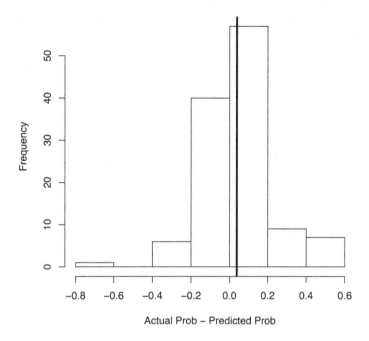

FIGURE 8.2
Histogram of the difference in the actual propensity of going for it and the predicted propensity at each yard line and to go distance. Vertical line at the mean difference between actual and predicted.

Figure 8.2 presents a histogram of the difference between the actual and predicted probability of going for it at each yard line and to go distance. It does not provide strong evidence of irrationality. In most cases the probabilities are the same or similar. However, there are cases where the model makes a strong prediction to go for it, but no NFL coach does. There are also cases where many NFL coaches do in fact go for it but the model does not predict that they should.

8.6 Discussion and Further Reading

The chapter uses game theory to solve the problem of estimating the value of punting the ball on 4th Down in American football. The estimation problem follows from the fact that coaches rarely go for it on 4th Down, and so there is no data. Using game theory and the GMM estimator, we can estimate the policy parameters using data from third downs. We can then use these estimates and the mixed strategy Nash equilibrium to simulate 4th Down. Our analysis suggests that NFL coaches punt the ball the appropriate number of times on 4th Down.

While GMM has become a standard technique, recent work suggests using moment *inequalities* generated from decision problems or games (Pakes et al., 2015).

9

Estimating Auction Models

9.1 Introduction

According to the travel presenter, Rick Steves, the Aalsmeer auction house is one the largest commercial buildings in the world. Royal Flora Holland, the owner of Aalsmeer, sold 12.5 billion plants and flowers in 2016 through its auction houses. But with $5.2 billion in auction sales, Royal Flora Holland is nowhere near the biggest auction house in the world.[1] That honor goes to Google. Google sold $47.6 billion in search ads using what the economist, Hal Varian, called the biggest auction in the world (Varian, 2007).[2] But while that is impressive, a single auction in 2015 almost beat Google's annual number. The US Federal Communication Commission's auction number 97 (AWS-3) raised $44.9 billion dollars.[3]

Auctions are used to sell and buy a large number of products. Governments use auctions to purchase everything from paper to police body cameras. The US Federal government uses auctions to sell oil drilling rights, FCC spectrum, 10 year bonds and timber access. You can sell and buy items from eBay.com using auctions.

Economists use auction data to determine the underlying valuations for the items. We may be interested in modeling what would happen if a different auction method is used. Economists working in antitrust want to determine the effect of a merger that reduces the number of independent bidders or determine if there is evidence of collusion.

The auctions at Aalsmeer are unique. The auction runs for a short amount of time with a "clock" clicking the price down as the auction continues. As the price falls, the first bidder to hit the button, wins, at whatever price the clock is at. A spokesman for Aalsmeer stated that because the price falls, it is called a **Dutch auction**. But actually, she got the causality backwards. Because the Dutch popularized these types of auctions for selling flowers, we call them **Dutch auctions**.

The auction style you may be most familiar with is called an **English auction**. In this auction, there is an auctioneer who often speaks very very

---

[1]https://www.royalfloraholland.com
[2]eMarketer.com, 7/26/16
[3]https://www.fcc.gov/auction/97/factsheet

fast and does a lot of pointing while bidders hold up paddles or make hand gestures. In English auctions, the last bidder wins and pays the price at which the bidding stops.

Economic analysis of auctions began with William Vickrey's seminal 1961 paper, *Counterspeculation, Auctions, and Competitive Sealed Bid Tenders.* Vickrey pointed out that Dutch auctions and **sealed bid auctions** are **strategically equivalent**. In a standard sealed bid auction each bidder submits a secret written bid. The auctioneer chooses the highest bid, and the bidder pays the number written down in her bid.

Vickrey characterized what a bidder should optimally bid in such an auction. He then showed that the same bidder should bid exactly the same amount in a Dutch auction. That is, in a Dutch auction, the bidder should wait until the price falls to the number written down, and then hit the button. Vickrey showed that while these two auctions formats are strategically equivalent, they are not strategically equivalent to an English auction.

Vickrey invented a new auction. In a **Vickrey auction**, each bidder writes down a bid like in a standard sealed bid auction and the winner is the person who writes down the highest bid. However, the winner pays the amount written down by the second highest bidder. Vickrey showed that his auction was strategically equivalent to an English auction.

Econometric analysis of auctions can be split into two steps. In the first step, we use standard statistical methods to estimate the statistical parameters. In the second step, we use the game theory to determine the **structural parameters**. We previously used this approach to estimate demand.

This chapter discusses two of the most important auction formats, sealed bid auctions and English auctions. It presents estimators for both. The sealed bid auction estimation is based on Guerre et al. (2000). The English auction analysis uses the order statistic approach of Athey and Haile (2002). In both cases it presents results for simulated data and analysis of timber auctions. The chapter tests whether loggers are bidding rationally in sealed bid auctions and whether loggers colluded in English auctions.

9.2   Sealed Bid Auctions

Sealed bid auctions are one of the most commonly used auction formats. These auctions are very prominent in procurement, both in government and the private sector. In a sealed bid auction, each bidder writes down her bid and submits it to the auctioneer. The auctioneer sorts the bids from highest to lowest (or lowest to highest if they are buying instead of selling). The winner is the highest bidder and she pays the amount she wrote down. This is called a **first price auction** because the price is determined by the highest bid or first price.

Vickrey pointed out that sealed bid auctions are strategically complicated. To see this, assume that a bidder's utility for an item is equal to their intrinsic value for the item less the price they pay for the item. For example, a logger bidding in a timber auction will earn profits from the logs less the price paid to the US Forestry service for access to the trees. If a logger bids an amount equal to her expected profits, then if she wins she will earn nothing from the logging. It is optimal for the logger to shade her bid down. The problem is that the more she shades down, the lower her chance of winning the auction. The bidder must calculate the trade off between the probability of winning the auction and the value of winning the auction.

9.2.1 Sealed Bid Model

Assumption 7. *Independent Private Values (IPV). Let $v_i \overset{iid}{\sim} F$ where v_i is the value of bidder i and F is the distribution function.*

Assumption 7 makes the exposition a lot simpler. It also seems to be a reasonable approximation for the problems considered. It states that a bidder's value for the item is unrelated to the values of the other bidders in the auction, except that they draw their valuation from the same distribution. It is a standard simplifying assumption in the auction literature. A contrasting assumption is called "common values." In a common values auction the item has the exact same value for everyone. Often it is assumed that while the bidders know that they all have the same value, they don't know exactly what that value is. This leads to an over-bidding problem called the "winner's curse."

The bidder maximizes her expected returns from the auction. Assume that the bidder gets 0 if she loses. If she wins, assume she gets her intrinsic value for the item less her bid.

$$\max_{b_{ij}} \quad \Pr(\text{win}|b_{ij})(v_{ij} - b_{ij}) \tag{9.1}$$

where b_{ij} represents the bid of bidder i in auction j, and v_{ij} represents the intrinsic value of the item for bidder i in auction j.

If we take first order conditions of Equation (9.1) then we get the following expression.

$$g(b_i|N)(v_i - b_i) - G(b_i|N) = 0 \tag{9.2}$$

Let $G(b_i|N)$ denote the probability that bidder i is the highest bidder with a bid of b_i, conditional on there being N bidders in the auction, and $g(b_i|N)$ is the derivative.

We can rearrange this formula to show how much the bidder should shade her bid.

$$b_i = v_i - \frac{G(b_i|N)}{g(b_i|N)} \tag{9.3}$$

The formula states that the bidder should bid her value, less a shading factor which is determined by how much a decrease in her bid reduces her probability of winning the auction.

It will be useful for our code to write the probability of winning the auction as a function of the bid distribution. Let $H(b)$ denote the distribution of bids in the auctions. Given Assumption 7, the probability of a particular bidder winning the auction is given by the following equation.

$$G(b_i|N) = H(b_i)^{N-1} \tag{9.4}$$

If there are two bidders in the auction, then the probability of winning is simply the probability that your bid is higher than the other bidder. If there are more than two bidders, it is the probability that your bid is higher than *all* the other bidders. The independent private values assumption implies it is the probability that each of the other bidders makes a bid less than yours, all multiplied together.

We can also determine the derivative of this function in terms of the bid.

$$g(b_i|N) = (N-1)h(b_i)H(b_i)^{N-2} \tag{9.5}$$

where h is the derivative of the bid distribution H.

9.2.2 Sealed Bid Simulation

In the simulated data, each bidder draws their value from a uniform distribution. Vickrey shows that the optimal bid in this auction is calculated using the following formula.

$$b_i = \frac{(N-1)v_i}{N} \tag{9.6}$$

Vickrey assumes that each bidder knows his own valuation, but only knows the distribution of valuations for other bidders in the auction. In game theory, this is called a game of **incomplete information**. We generally assume that the outcome of such games is a **Bayes Nash equilibrium**.[4]

Remember a game has three parts, a set of players, a set of strategies and a set of payoffs. In the case of auctions, the players are the bidders in the auction. The strategies are the bids. Actually, that is not quite correct. In games of incomplete information, the players do not observe the actions of the other players. When bidding, a bidder does not know what the other bidders are bidding. Instead, it is assumed that bidders know the function that maps from valuations to bids. In Vickrey's game, bidders know the function represented by Equation (9.6). They also know their own valuation. The payoffs are the expected value of the auction accounting for the probability of winning, the intrinsic value of the item and the amount bid.

The uniform distribution simplifies the problem, which is why it is used. In each simulated auction, there are different numbers of simulated bidders.

---

[4]This is a Nash equilibrium where it is assumed that players use Bayes' rule to update their information given the equilibrium strategies. Because of the IPV assumption, there is no information provided by the other bidders. This is not the case in common values auctions.

```
> set.seed(123456789)
> M <- 1000   # number of simulated auctions.
> data1 <- matrix(NA,M,12)
> for (i in 1:M) {
+    N <- round(runif(1, min=2,max=10)) # number of bidders.
+    v <- runif(N) # valuations, uniform distribution.
+    b <- (N - 1)*v/N # bid function
+    p <- max(b)  # auction price
+    x <- rep(NA,10)
+    x[1:N] <- b  # bid data
+    data1[i,1] <- N
+    data1[i,2] <- p
+    data1[i,3:12] <- x
+ }
> colnames(data1) <- c("Num","Price","Bid1",
+                        "Bid2","Bid3","Bid4",
+                        "Bid5","Bid6","Bid7",
+                        "Bid8","Bid9","Bid10")
> data1 <- as.data.frame(data1)
```

The simulation creates a data set with 1,000 auctions. In each auction, there is between 2 and 10 bidders. Note that the bidders are not listed in order.

9.2.3   Sealed Bid Estimator

The estimator uses Equation (9.3) to back out values from observed bids. To do this, we calculate the probability of winning the auction conditional on the number of bidders. It should be straightforward to determine from this data. Once we have this function, we use the formula to determine the bidder's valuation from their bid.

The first step is to estimate the bid distribution.

$$\hat{H}(b) = \frac{1}{N} \sum_{i=1}^{N} \mathbb{1}(b_i < b) \tag{9.7}$$

The **non-parametric estimate** of the distribution function, $H(b)$, is the fraction of bids that are below some value b.[5]

The second step is to estimate the derivative of the bid distribution. This can be calculated numerically for some given "small" number, ϵ.[6]

$$\hat{h}(b) = \frac{\hat{H}(b + \epsilon) \quad \hat{H}(b - \epsilon)}{2\epsilon} \tag{9.8}$$

---

[5] A non-parametric estimator makes no parametric assumptions.
[6] This is a **finite difference estimator**.

If there are two bidders, Equation (9.3) determines the valuation for each bidder.

$$\hat{v}_i = b_i + \frac{\hat{H}(b_i)}{\hat{h}(b_i)} \tag{9.9}$$

where $i \in \{1, 2\}$.

9.2.4   Sealed Bid Estimator in R

The estimator backs out the valuation distribution from the distribution of bids. It limits the data to only those auctions with two bidders. In this special case, the probability of winning is just given by the distribution of bids.[7] In the code the term "epsilon" stands for the Greek letter, ϵ, and refers to a "small" number.

```
> f_sealed_2bid <- function(bids, epsilon=0.5) {
+    # epsilon for "small" number for finite difference method
+    # of taking numerical derivatives.
+    values <- rep(NA,length(bids))
+    for (i in 1:length(bids)) {
+      H_hat <- mean(bids < bids[i])
+      # bid probability distribution
+      h_hat <- (mean(bids < bids[i] + epsilon) -
+        mean(bids < bids[i] - epsilon))/(2*epsilon)
+      # bid density
+      values[i] <- bids[i] + H_hat/h_hat
+    }
+    return(values)
+ }
> bids <- data1[data1$Num==2,3:4]
> bids <- as.vector(as.matrix(bids)) # all the bids
> values <- f_sealed_2bid(bids)
```

It is straightforward to calculate the probability of winning, as this is the probability the other bidder bids less. Given IPV, this is just the cumulative probability for a particular bid. Calculating the density is slightly more complicated. We can approximate this derivative numerically by looking at the change in the probability for a "small" change in the bids.[8] The value is calculated using Equation (9.3).

Figure 9.1 shows that the bids are significantly shaded from the true values, particularly for very high valuations. The figure presents the density functions for bids and derived valuations from the two-person auctions. The true density of valuations lies at 0.5 and goes from 0 to 1. Here the estimated density is a

---

[7]The probability of winning is the probability that your bid is higher than the other bidders in the auction.

[8]This is an example of using finite differences to calculate numerical derivatives.

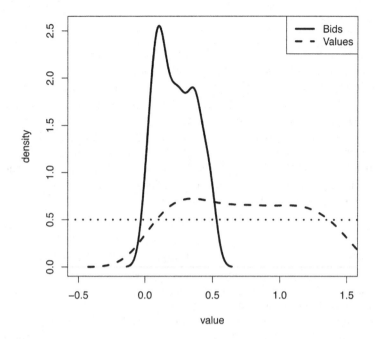

FIGURE 9.1
Plot of the density function for bids and values from 2 person auctions. The true distribution of valuations has a density of 0.5 from 0 to 1.

little higher and goes over its bounds. However, part of the reason may be the method we are using to represent the density in the figure.[9]

9.3 English Auctions

The auction format that people are most familiar with is the English auction. These auctions are used to sell cattle, antiques, collector stamps and houses (in Australia). In the 1970s they were also the standard format used by the US Forestry Service to sell timber access (Aryal et al., 2018).

Vickrey showed that English auctions are strategically very simple. Imagine a bidder hires an expert auction consultant to help them bid in an English auction.

- Expert: "What is your value for the item?"

---

[9]The kernel density method assumes the distribution can be approximated as a mixture of normal distributions.

- Bidder: "$2,300"

- Expert: "Bid up to $2,300 and then stop."

In sealed bid auctions there is an optimal trade-off between winning and profiting from the auction. In English auctions there is no such trade-off. In econ-speak the high bidder is pivotal in determining the price in sealed bid auctions, but is not pivotal in English auctions.

English auctions are **second price auctions**. The price in an English auction is determined by the highest *losing* bid. That is, it is determined by the second price. In English auctions the bidding continues until the second to the last bidder drops out. Once there is only one bidder, everything stops.

The optimal bid for bidder i is to bid her value.

$$b_i = v_i \qquad (9.10)$$

Equation (9.10) suggests that empirical analysis of English auctions is a lot simpler than for sealed bid auctions. If only that were so! To be clear, the "bid" in Equation (9.10) means the strategy described by the expert. In the data we do not necessarily observe this strategy.

If we could observe all the bid strategies in the auction, then we would have an estimate of the value distribution. But that tends to be the problem. Depending on the context, not all active bidders in the auction may actually be observed making a bid. In addition, if the price jumps during the auction we may not have a good idea when bidders stopped bidding (Haile and Tamer, 2003).

Athey and Haile (2002) provide a solution. They point out that the price in an English auction has a straightforward interpretation. When valuations follow Assumption 7, the price is the second highest valuation of the people who bid in the auction. Consider if the price is lower than the second highest valuation. How could that be? Why did one of the bidders exit the auction at a price lower than her valuation? Consider if the price is higher than the second highest valuation. How could that be? Why would a bidder bid more than her valuation?

If the price is equal to the second highest valuation, then it is a particular order statistic of the value distribution. Athey and Haile (2002) show how the observed distributions of an order statistic uniquely determine the value distribution.

9.3.1   Order Statistics

To understand how order statistics work, consider the problem of determining the distribution of heights of players in the WNBA. The obvious way to do it is to take a data set on player heights and calculate the distribution. A less obvious way is to use order statistics.

In this method, data is taken from a random sample of teams, where for each team, the height of the tallest player is measured. Assume each team has 10 players on the roster and you know the height of the tallest, say the center. This is enough information to estimate the distribution of heights in the WBNA. We can use the math of order statistics and the fact that we know both the height of the tallest and we know that 9 other players are shorter. In this case we are using the tallest, but you can do the same method with the shortest or the second tallest, etc.

The price is more or less equal to the second highest valuation of the bidders in the auction.[10] The probability of the second highest of N valuations is equal to some value b which is given by the following formula:

$$\Pr(b^{(N-1):N} = b) = N(N-1)F(b)^{N-2}f(b)(1 - F(b)) \qquad (9.11)$$

The order statistic notation for the second highest bid of N is $b^{(N-1):N}$. We can parse this equation from back to front. It states that the probability of seeing a price equal to b is the probability that one bidder has a value greater than b. This is the winner of the auction and this probability is given by $1 - F(b)$, where $F(b)$ is the cumulative probability of a bidder's valuation less than b. This probability is multiplied by the probability that there is exactly one bidder with a valuation of b. This is the second highest bidder who is assumed to bid her value. This is represented by the density function $f(b)$. These two values are multiplied by the probability that the remaining bidders have valuations less than b. If there are N bidders in the auction then $N - 2$ of them have valuations less than the price. The probability of this occurring is $F(b)^{N-2}$. Lastly, the labeling of the bidders is irrelevant so there are $\frac{N!}{1!(N-2)!} = N(N-1)$ possible combinations. If the auction has two bidders, then the probability of observing a price p is $2f(p)(1 - F(p))$.

The question raised by Athey and Haile (2002) is whether we can use this formula to determine F. Can we use the order statistic formula of the distribution of prices to uncover the underlying distribution of valuations? Yes.

9.3.2 Identifying the Value Distribution

Let's say we observe a two-bidder auction with a price equal to the lowest possible valuation for the item; call that v_0. Actually, it is a lot easier to think about the case where the price is slightly above the lowest possible value. Say that the price is less than $v_1 = v_0 + \epsilon$, where ϵ is a "small" number. What do we know? We know that one bidder has that very low valuation, which occurs with probability equal to $F(v_1)$. What about the other bidder? The other bidder may also have a value equal to the lowest valuation or they may have a higher valuation. That is, their value for the item could be anything. The probability of value lying between the highest and lowest possible value is 1. So $\Pr(p \le v_1) = 2 \times 1 \times F(v_1)$. Note that either bidder could be the high

---

[10]Officially, the price may be a small increment above the bid of the second highest bidder. We will ignore this possibility.

bidder. There are 2 possibilities, so we must multiply by 2. As the probability of a price less than v_1 is observed in the data, we can rearrange things to get the initial probability, $F(v_1) = \Pr(p \le v_1)/2$.

Now take another value, $v_2 = v_1 + \epsilon$. The probability of observing a price between v_1 and v_2 is as follows.

$$\Pr(p \in (v_1, v_2]) = 2(F(v_2) - F(v_1))(1 - F(v_1)) \tag{9.12}$$

It is the probability of seeing one bidder with a value between v_2 and v_1 and the second bidder with a value greater than v_1. Again, the two bidders can be ordered in two ways.

We can solve $F(v_2)$ using Equation (9.12). We observe the quantity on the left-hand side and we previously calculated $F(v_1)$.

For a finite subset of the valuations we can use this iterative method to calculate the whole distribution. For this to work, each bidder's valuation is assumed to be independent of the other bidders and comes from the same distribution of valuations (Assumption 7).

9.3.3 English Auction Estimator

The non-parametric estimator of the distribution follows the logic above.

The initial step determines the probability at the minimum value,

$$\hat{F}(v_1) = \frac{\sum_{j=1}^{M} \mathbb{1}(p_j \le v_1)}{2M} \tag{9.13}$$

where there are M auctions and p_j is the price in auction j.

To this initial condition, we can add an iteration equation.

$$\hat{F}(v_k) = \frac{\sum_{j=1}^{M} \mathbb{1}(v_k < p_j \le v_{k+1})}{2M(1 - \hat{F}(v_{k-1}))} + \hat{F}(v_{k-1}) \tag{9.14}$$

These equations are then used to determine the distribution of the valuations.

9.3.4 English Auction Estimator in R

We can estimate the distribution function non-parametrically by approximating it at $K = 100$ points evenly distributed across the range of observed values. The estimator is based on Equations (9.13) and (9.14).

```
> f_English_2bid <- function(price, K=100, epsilon=1e-8) {
+    # K number of finite values.
+    # epsilon small number for getting the probabilities
+    # calculated correctly.
+    min1 <- min(price)
+    max1 <- max(price)
```

```
+    diff1 <- (max1 - min1)/K
+    Fv <- matrix(NA,K,2)
+    min_temp <- min1 - epsilon
+    max_temp <- min_temp + diff1
+    # determines the boundaries of the cell.
+    Fv[1,1] <- (min_temp + max_temp)/2
+    gp <- mean(price > min_temp & price < max_temp)
+    # price probability
+    Fv[1,2] <- gp/2  # initial probability
+    for (k in 2:K) {
+       min_temp <- max_temp - epsilon
+       max_temp <- min_temp + diff1
+       Fv[k,1] <- (min_temp + max_temp)/2
+       gp <- mean(price > min_temp & price < max_temp)
+       Fv[k,2] <- gp/(2*(1 - Fv[k-1,2])) + Fv[k-1,2]
+       # cumulative probability
+    }
+    return(Fv)
+ }
```

Consider simulated data from an English auction. The data set provides the price and the number of bidders in each auction.

```
> M <- 10000
> data2 <- matrix(NA,M,2)
> for (i in 1:M) {
+    N <- round(runif(1, min=2,max=10))
+    v <- rnorm(N) # normally distributed values
+    b <- v  # bid
+    p <- -sort(-b)[2]   # auction price
+    data2[i,1] <- N
+    data2[i,2] <- p
+ }
> colnames(data2) <- c("Num","Price")
> data2 <- as.data.frame(data2)
```

Given this data we can determine the value distribution for the two-bidder auctions.

```
> price <- data2$Price[data2$Num==2]
> # restrics the data to auctions w/ 2 bidders.
> Fv <- f_English_2bid(price)
```

Figure 9.2 shows that the method does a pretty good job at estimating the underlying distribution. The true distribution has a median of 0, which means that the curve should go through the point where the two dotted lines cross.

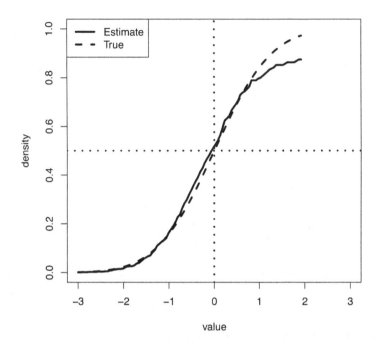

FIGURE 9.2
Plot of the distribution function for values from two person English auctions.
The true distribution of valuations is a standard normal distribution.

However, the estimate of the upper tail is not good. The true distribution has a
much thicker upper tail than the estimated distribution. How does the estimate
change if you use auctions with a larger number of bidders? Remember to
change the formula appropriately.

9.4   Are Loggers Rational?

In the 1970s, the US Forest Service conducted an interesting experiment. It
introduced sealed bid auctions in 1977. Previous to that, most US Forest
Service auctions had been English auctions.[11] In 1977, the service mixed
between auction formats. As discussed above, bidding in sealed bid auctions

---

[11] You may think of this as just some academic question. But the US Senator for Idaho,
Senator Church, was not happy with the decision. "In fact, there is a growing body of
evidence that shows that serious economic dislocations may already be occurring as a result
of the sealed bid requirement." See *Congressional Record* September 14 1977, p. 29223.

is strategically a lot more complicated than bidding in English auctions. In the latter, the bidder simply bids her value. In the former, she must trade off between bidding higher and increasing the likelihood of winning against paying more if she does win.

Because of the experiment, we can test whether the loggers in the sealed bid auctions bid consistently with their actions in the English auctions. Our test involves estimating the underlying value distribution using bid data from sealed bid auctions, and comparing that to an estimate of the underlying value distribution using price data from English auctions.

9.4.1   Timber Data

The data used here is from the US Forest Service downloaded from Phil Haile's website.[12]

In order to estimate the distributions of bids and valuations it is helpful to "normalize" them so that we are comparing apples to apples. The standard method is to use a log function of the bid amount and run OLS on various characteristics of the auction including the number of acres bid on, the estimated value of the timber, access costs and characteristics of the forest and species (Haile et al., 2006).[13]

```
> x <- read.csv("auctions.csv", as.is = TRUE)
> lm1 <- lm(log_amount ~ as.factor(Salvage) + Acres +
+              Sale.Size + log_value + Haul +
+              Road.Construction + as.factor(Species) +
+              as.factor(Region) + as.factor(Forest) +
+              as.factor(District), data=x)
> # as.factor creates a dummy variable for each entry under the
> # variable name.  For example, it will have a dummy for each
> # species in the data.
> x$norm_bid <- NA
> x$norm_bid[-lm1$na.action] <- lm1$residuals
> # lm object includes "residuals" term which is the difference
> # between the model estimate and the observed outcome.
> # na.action accounts for the fact that lm drops
> # missing variables (NAs)
```

In general, we are looking for a normal-like distribution. Figure 9.3 presents the histogram of the normalized log bids. It is not required that the distribution

---

[12]http://www.econ.yale.edu/~pah29/timber/timber.htm. The version used here is available from here: https://sites.google.com/view/microeconometricswithr/table-of-contents.

[13]Baldwin et al. (1997) discuss the importance of various observable characteristics of timber auctions.

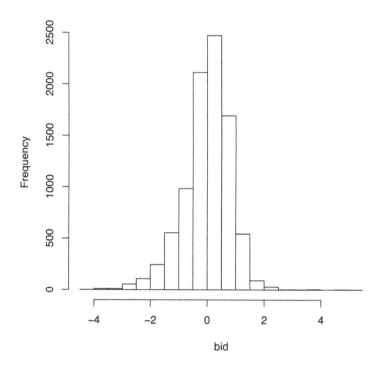

FIGURE 9.3
Histogram of normalized bid residual for US Forest Service auctions from 1977.

be normal, but if the distribution is quite different from normal, you should think about why that may be. Does this distribution look normal?[14]

9.4.2 Sealed Bid Auctions

In order to simplify things we will limit the analysis to two-bidder auctions. In the data, sealed bid auctions are denoted "S".

```
> y <- x[x$num_bidders==2 & x$Method.of.Sale=="S",]
> bids <- y$norm_bid
> values <- f_sealed_2bid(bids)
> summary(bids)
```

---

[14]It is approximately normal, but it is skewed somewhat to lower values. This may be due to low bids in the English auction. How does the distribution look if only sealed bids are graphed?

```
   Min.  1st Qu.   Median     Mean 3rd Qu.     Max.
-4.0262 -0.7987  -0.1918  -0.2419   0.3437   5.0647
```

```
> summary(values)
```

```
   Min.   1st Qu.    Median     Mean  3rd Qu.      Max.
-4.0262  -0.0218    0.9185   4.8704   2.3108  860.0647
```

Using the same method that we used above, it is possible to back out an estimate of the value distribution from the bids in the data. We see that comparing the valuations to the bids, the bids are significantly shaded particularly for higher valuations.

9.4.3 English Auctions

Again consider two-bidder auctions. The English auctions are denoted "A".

```
> y <- x[x$num_bidders==2 & x$Method.of.Sale=="A",]
> price <- y[y$Rank==2,]$norm_bid
> Fv <- f_English_2bid(price)
```

We can back out the value distribution assuming that the price is the second highest bid, the second highest order statistic.

9.4.4 Comparing Estimates

Figure 9.4 shows that there is not a whole lot of difference between the estimate of the distribution of valuations from sealed bid auctions and English auctions. The two distributions of valuations from the sealed bid auctions and English auctions lie fairly close to each other, particularly for lower values. This suggests loggers are bidding rationally. That said, at higher values, the two distributions diverge. The value distribution from the sealed bid auctions suggests that valuations are higher than the estimate from the English auctions. What else may explain this divergence?

9.5 Are Loggers Colluding?

Is there evidence that bidders in these auctions are colluding? Above, we find mixed evidence that bidders in timber auctions are behaving irrationally. Now we can ask whether they are behaving competitively.

Given the price is an order statistic, prices should be increasing in the number of bidders. Figure 9.5 presents a plot of the number of bidders in an English auction against the price. The figure suggests that this relationship

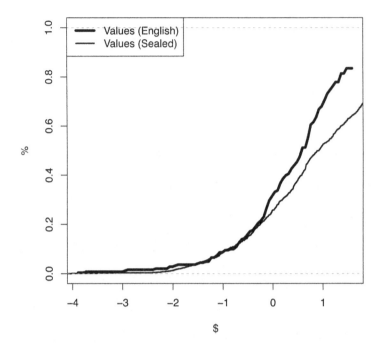

FIGURE 9.4
Comparison of estimated distributions from two bidder English and sealed bid auctions.

is not clear. One explanation is that bidders are colluding. It could be that bidders in larger auctions are underbidding. Of course, there may be many other reasons for the observed relationship.

9.5.1   A Test of Collusion

Consider the following test of collusion. Using large English auctions we can estimate the distribution of valuations. Under the prevailing assumptions, this estimate should be the same as for two-bidder auctions. If the estimate from the large auctions suggests valuations are much lower than for two-bidder auctions, this suggests collusion.

In particular, if the inferred valuations in these larger auctions look much like auctions with fewer bidders. That is, bidders may behave "as if" there are actually fewer bidders in the auction. For example, if there is an active **bid-ring**, bidders may have a mechanism for determining who will win the auction and how the losers may be compensated for not bidding.[15] In an English auction,

---

[15] Asker (2010) presents a detailed account of a bid-ring in stamp auctions.

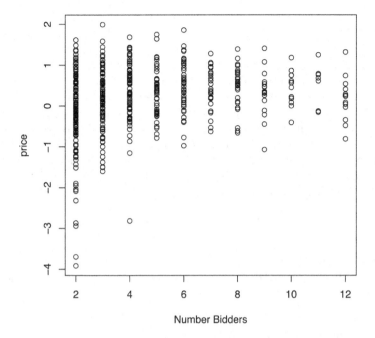

FIGURE 9.5
Plot of prices against number of bidders.

it is simple to enforce a collusive agreement because members of the bid-ring can bid in the auction where their actions are observed.

We can estimate the distribution under the assumption that there are two bidders in the auction and the assumption that there are three bidders in the auction. If these estimates are consistent with the results from the actual two bidder auctions, then we have an estimate of the bid-ring size. Consider an auction with six bidders. If three of them are members of a bid-ring, then those three will agree on who should bid from the ring. Only one of the members of the bid-ring will bid their value. In the data, this will look like there are actually four bidders in the auction.[16]

9.5.2 "Large" English Auctions

Consider the case where there are six bidders in the auction. From above, the order statistic formula for this case is as follows.

$$\Pr(b^{5:6} = b) = 30F(b)^4 f(b)(1 - F(b)) \tag{9.15}$$

---

[16] In the bid-ring mechanism discussed in Asker (2010), the collusion actually leads to higher prices in the main auction.

As above, order statistics are used to determine the underlying value distribution (F); however in this case it is a little more complicated to determine the starting value.

Think about the situation where the price in a 6 bidder auction is observed at the minimum valuation. What do we know? As before, one bidder may have a value equal to the minimum or a value above the minimum. That is, their value could be anything. The probability of a valuation lying between the minimum and maximum value is 1. We also know that the five other bidders had valuations at the minimum. If not, one of them would have bid more and the price would have been higher. As there are six bidders, there are six different bidders that could have had the highest valuation. This reasoning gives the following formula for the starting value.

$$\Pr(b^{5:6} < v_1) = 6F(v_1)^5 \qquad (9.16)$$

Rearranging, we have $F(v_1) = \left(\frac{\Pr(p \le v_1)}{6}\right)^{\frac{1}{5}}$.

Given this formula we can use the same method as above to solve for the distribution of valuations.

9.5.3   Large English Auction Estimator

Again we can estimate the value distribution by using an iterative process. In this case we have the following estimators.

$$\hat{F}(v_1) = \left(\frac{\sum_{j=1}^{M} \mathbb{1}(p_j < v_1)}{6M}\right)^{\frac{1}{5}} \qquad (9.17)$$

and

$$\hat{F}(v_k) = \frac{\sum_{j=1}^{M} \mathbb{1}(v_k < p_j < v_{k+1})}{30M\hat{F}(v_{k-1})^4(1 - \hat{F}(v_{k-1}))} + \hat{F}(v_{k-1}) \qquad (9.18)$$

The other functions are as defined in the previous section.

We can also solve for the implied distribution under the assumption that there are three bidders and under the assumption that there are two bidders.[17] Note in each auction there are at least six bidders.[18]

9.5.4   Large English Auction Estimator in R

We can adjust the estimator above to allow any number of bidders, N.

---

[17]See Equation (9.11) for the other cases.

[18]For simplicity it is assumed that all of these auctions have six bidders. Once there are a large enough number of bidders in the auction, prices do not really change with more bidders. In fact, these methods may not work as the number of bidders gets large (Deltas, 2004).

```
> f_English_Nbid <- function(price, N, K=100, epsilon=1e-8) {
+    min1 <- min(price)
+    max1 <- max(price)
+    diff1 <- (max1 - min1)/K
+    Fv <- matrix(NA,K,2)
+    min_temp <- min1 - epsilon
+    max_temp <- min_temp + diff1
+    Fv[1,1] <- (min_temp + max_temp)/2
+    gp <- mean(price > min_temp & price < max_temp)
+    Fv[1,2] <- (gp/N)^(1/(N-1))
+    for (k in 2:K) {
+      min_temp <- max_temp - epsilon
+      max_temp <- min_temp + diff1
+      Fv[k,1] <- (min_temp + max_temp)/2
+      gp <- mean(price > min_temp & price < max_temp)
+      Fv[k,2] <-
+        gp/(N*(N-1)*(Fv[k-1,2]^(N-2))*(1 - Fv[k-1,2])) +
+        Fv[k-1,2]
+    }
+    return(Fv)
+ }
> y_6 <- x[x$num_bidders > 5 & x$Method.of.Sale=="A",]
> # English auctions with at least 6 bidders.
> price_6 <- y_6$norm_bid[y_6$Rank==2]
> Fv_6 <- f_English_Nbid(price_6, N=6)
> Fv_3 <- f_English_Nbid(price_6, N=3)
> Fv_2 <- f_English_2bid(price_6)
```

9.5.5 Evidence of Collusion

Figure 9.6 suggests that there is in fact collusion in these auctions! Assuming there are six bidders in the auction implies that valuations are much lower than we estimated for two-bidder auctions from both English auctions and sealed bid auctions. In the chart, the distribution function is shifted to the left, meaning there is greater probability of lower valuations.

If we assume the estimates from the two-bidder auctions are the truth, then we can determine the size of the bid-ring. Estimates assuming there are three bidders and two bidders lie above and below the true value, respectively. This suggests that bidders are behaving as if there are between two and three bidders in the auction. This implies that the bid-ring has between three and four bidders in each auction.

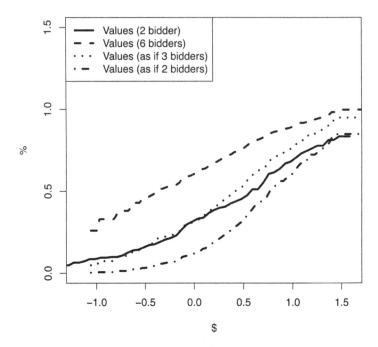

FIGURE 9.6
Comparison of estimated distribution of valuations from English auctions with
at least 6 bidders. These estimates are compared to the estimate from 2 bidder
auctions.

These results are suggestive of an active bid-ring in these auctions in 1977.
It turns out that this was of real concern. In 1977, the United States Senate
conducted hearings into collusion in these auctions. In fact, this may be why
the US Forestry Service looked into changing to sealed bid auctions. The US
Department of Justice also brought cases against loggers and millers (Baldwin
et al., 1997). Alternative empirical approaches have also found evidence of
collusion in these auctions, including Baldwin et al. (1997) and Athey et al.
(2011).

9.5.6   Large Sealed Bid Auction Estimator

One concern with the previous analysis is that there may be something special
about auctions with a relatively large number of bidders. It may that there are
unobservable characteristics of the auctions that are leading to the observed
differences. To check this possibility we can compare our estimate of the value
distribution from large sealed bid auctions.

For this case the probability of winning the auction is slightly different than the simple case above.

$$\hat{G}(b|6) = \hat{H}(b)^5 \qquad (9.19)$$

and

$$\hat{g}(b|6) = 5\hat{h}(b)\hat{H}(b)^4 \qquad (9.20)$$

where $\hat{H}(b)$ and $\hat{h}(b)$ are defined above.

9.5.7 Large Sealed Bid Auction Estimator in R

```
> f_sealed_Nbid <- function(bids, N, epsilon=0.5) {
+    values <- rep(NA,length(bids))
+    for (i in 1:length(bids)) {
+      H_hat <- mean(bids < bids[i])
+      h_hat <- (mean(bids < bids[i] + epsilon) -
+        mean(bids < bids[i] - epsilon))/(2*epsilon)
+      G_hat <- H_hat^(N-1)
+      g_hat <- (N-1)*h_hat*H_hat^(N-2)
+      values[i] <- bids[i] + G_hat/g_hat
+    }
+    return(values)
+ }
> y_6 <- x[x$num_bidders > 5 & x$Method.of.Sale=="S",]
> bids_6 <- y_6$norm_bid
> values_6 <- f_sealed_Nbid(bids_6, N=6,epsilon=3)
```

As for two person auctions, the first step is to estimate each bidder's valuation.

The results presented in Figure 9.7 provide additional evidence that there is collusion in the large English auctions. The bidding behavior in the larger sealed bid auctions is consistent with both bidding behavior in small sealed bid auctions and small English auctions.

9.6 Discussion and Further Reading

Economic analysis of auctions began with Vickrey's 1961 paper. Vickrey used game theory to analyze sealed bid auctions, Dutch auctions and English auctions. Vickrey also derived a new auction, the sealed bid second price auction. In 2020, the field was honored with Nobel prizes awarded to Paul Milgrom and Robert Wilson. Empirical analysis of auctions starts with the game theory.

The chapter considers two of the most important auction mechanisms,

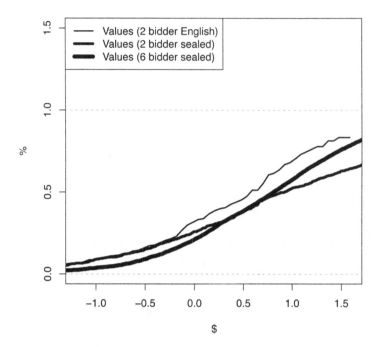

FIGURE 9.7
Comparison of estimated distribution of valuations from sealed bid auctions
with at least six bidders. These estimates are compared to the estimate from
two-bidder auctions.

sealed bid auctions and English auctions. The sealed bid auctions are analyzed
using the two-step procedure of Guerre et al. (2000). The first step uses non-
parametric methods to estimate the bid distribution. The second step uses the
Nash equilibrium to back out the value distribution.

While we observe all the bids in the sealed bid auctions, we generally only
observe the high bids in English auctions. The chapter uses the order statistic
approach of Athey and Haile (2002) to estimate the value distribution from
these auctions.

Empirical analysis of auctions has changed dramatically over the last twenty
years. This section just grazed the surface of the different issues that arise in
auctions. One big issue not discussed here is unobserved auction heterogeneity.
There may be systematic differences between the auctions that are observed by
the bidders but not observed by the econometrician. The most important paper
written on this question was written by a former office mate of mine, Elena
Krasnokutskaya, (Krasnokutskaya, 2011). Krasnokutskaya uses an approach
similar to the mixture model approach discussed in Chapter 12. A recent review

by Phil Haile and Yuichi Kitamura is an excellent starting point (Haile and Kitamura, 2019).

Baldwin et al. (1997) and Athey et al. (2011) analyze collusion in US timber auctions. Aryal et al. (2018) uses US timber auctions to measure how decision makers account for risk and uncertainty.

Part III

Repeated Measurement

10

Panel Data

10.1 Introduction

The first part of the book considers the value of **exogenous** changes in variables to help predict the effect of a policy change. The second part of the book illustrates how economic theory can be used explicitly to predict the effect. This part argues that repeated measurement provides valuable information for predicting the effect of a policy change.

Repeated measurement of the same individual allows us to infer the individual treatment effect and uncover unobserved characteristics of the individual. This is of particular interest with **panel data** sets. In microeconomic panel data sets, such as the National Longitudinal Survey of Youth, we observe a large number of individuals across many years. This chapter works through the standard approaches of **difference in difference** and **fixed effects**. These methods are useful when a subset of individuals observed in the data, face differing levels of exposure to the policy over time.

10.2 First Differences

Consider data generated according to Figure 10.1. We are interested in the causal effect of X on Y which is denoted b_i. Note that this effect may vary with the individual denoted i. Now consider that we observe two different X's for the same individual. Say two different education levels. By observing the same individual with two different education levels we can estimate the individual effect on the outcome of interest. We can measure the difference in income associated with the increase in education. We can measure the **causal effect** of education on income for each individual.

However, if we observe the same individual with two different education levels then we observe the same individual at two different *times*. Unobserved characteristics of the individual may also change between the two time periods. There is both an effect of education and of Time on Y. The time effect is denoted d_i in the graph.

Figure 10.1 shows that the relationship between X and Y is **confounded**.

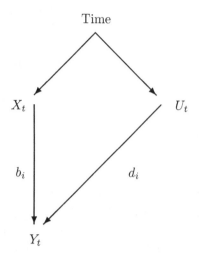

FIGURE 10.1
Confounded time graph.

In the terminology of Judea Pearl, there is a **backdoor relationship** that works through Time. That is, you can follow a line directly from X to Y and follow a line "backwards" from X to $Time$ and $Time$ to U and then Y. If the effect of $Time$ on X and U is 1, then the regression of X on Y is equal to $b_i + d_i$. In order to estimate b_i we need to find a way to estimate d_i.

This section estimates the policy effect by taking first differences in the outcome and the policy variables.

10.2.1 First Difference Model

Consider that we observe an outcome of interest for individual i at two time periods $t \in \{1, 2\}$.

$$y_{it} = a + bx_{it} + dv_{it} + \epsilon_{it} \qquad (10.1)$$

where the outcome of interest y_{it} is determined by the policy variable x_{it} *and* some unobserved time effect v_{it}.

In Chapter 2 we accounted for this additional variable by including it in the regression. Here we can account for the unobserved time effect by taking first differences.

$$y_{i2} - y_{i1} = b(x_{i2} - x_{i1}) + d(v_{i2} - v_{i1}) + \epsilon_{i2} - \epsilon_{i1} \qquad (10.2)$$

If we assume that the time effect is 1, then we can add an intercept to measure the time effect.

10.2.2 Simulated Panel Data

Consider the simulated panel data set created below. There are 1,000 simulated individuals observed over two time periods. Note that x is increasing, at least weakly, over time. What do you get if you run OLS of x on y?[1]

```
> set.seed(123456789)
> N <- 1000   # 1000 individuals
> T <- 2      # 2 time periods
> a <- -2
> b <- 3
> d <- -4
> u <- c(5,6)  # unobserved characteristic in the two periods.
> x1 <- runif(N)
> x2 <- x1 + runif(N)   # change in X over time.
> x <- as.matrix(rbind(x1,x2))   # creates a matrix T x N.
> e <- matrix(rnorm(N*T),nrow = T)
> # additional unobserved characteristic
> y <- a + b*x + d*u + e
> # matrix with T rows and N columns.
```

10.2.3 OLS Estimation of First Differences

The simulated data allows us to observe x change for the same individual. Given that we are interested in the effect of this change in x we could look at first differences. That is we could look at regressing the *change* in x on the *change* in y.

```
> diffy <- y[2,] - y[1,]
> diffx <- x[2,] - x[1,]
> lm1 <- lm(diffy ~ diffx)
```

Table 10.1 presents the results of the first difference approach. It does a pretty good job at getting at the true value of $b = 3$ if the intercept term is used. The intercept captures the time effect and is close to $d = -4$.

	Estimate	Std. Error	t value	Pr(>|t|)
(Intercept)	-3.7710	0.0901	-41.84	0.0000
diffx	2.7481	0.1542	17.83	0.0000

TABLE 10.1
OLS of first differences.

---

[1]You need to transform both x and y from matrices to vectors in order to run the regression.

10.3   Difference in Difference

Consider a case where we observe two types of individuals in the data. For example, let it be the case that for all individuals in the first time period we observe them with only a high school education. Then for the second time period, say ten years later, we observe two different groups. The first group has a college education. We call this the **treated group**, while the second group remains with a high school education. We call these people the **non-treated group**. For this second group, we can estimate the impact of time on the outcome of interest. For this group we can estimate the impact of the ten years on income. Therefore, if the effect of time is the *same* for both the **treated group** and the **non-treated group** we can disentangle the effect of time and the policy variable. We can disentangle the effect of the ten years and attending college on income. Of course, it is not at all obvious that the effect of time would be the same for both groups.

The section presents the difference in difference estimator and illustrates the estimator with simulated data.

10.3.1   Difference in Difference Estimator

In algebra, we have the following derivation of the difference in difference estimator. Assume that the observed outcome of interest for individual i in time t is determined by the treatment according to the following function.

$$y_{it}(x_{it}) = a + bx_{it} + dv_t + \epsilon_{it} \tag{10.3}$$

where $x_{it} \in \{0,1\}$ denotes if the individual receives the treatment. The equation says that individual i's outcome in time t is determined by both whether or not she has received the treatment and an unobserved term $v_t \in \{0,1\}$ that changes with time. There is also another unobserved term ϵ_{it}. We will assume that this second unobserved term has a mean of 0 and is independent of everything else that is going on. It is "noise."

Consider observing outcomes y_{it} for a person in the treated group in two different periods ($t \in \{0,1\}$). The first period is prior to the individual receiving the treatment. Say prior to going to college. The second period is after the individual receives the treatment and the person goes to college. From above, if we compare the expected outcomes from the two cases we get the following result.

$$\mathbb{E}(y_{i1}|x_{i1}=1, v_1=1) - \mathbb{E}(y_{i0}|x_{i0}=0, v_0=0) = b + d \tag{10.4}$$

where y_{i1} is the outcome observed in time period 1 for individual i and x_{i1} indicates the treatment received by individual i in period 1.

For an individual in the non-treated group, we have the following equation.

$$\mathbb{E}(y_{i1}|x_{i1}=0, v_1=1) - \mathbb{E}(y_{i1}|x_{i0}=0, v_0=0) = d \tag{10.5}$$

For this group, the change in the time period does not change the treatment that they receive. For this group, the only change in outcomes is due to time.

Now we can compare the two differences. Equation (10.4) considers two conditional expectations. The first conditions on the treatment level equal to 1. This only occurs to individuals that are in the treated group for the time period in which they are treated. This is the group that attends college. Equation (10.5) gives the difference in the conditional expectation for the non-treated group, conditional on the two time periods. Thus, we can estimate b by taking the second difference from the first difference.

The analog estimator is as follows.

$$\beta_{did} = \frac{1}{N}\left(\sum_{i=1}^{N} \mathbb{1}(x_{i1} = 1)(y_{i1} - y_{i0}) - \mathbb{1}(x_{i1} = 0)(y_{i1} - y_{i0})\right) \qquad (10.6)$$

where $\mathbb{1}()$ is an indicator function; it is 1 if the statement in the parentheses is true and 0 otherwise. The difference in difference estimator is the average difference in outcomes for the treated less the average difference in outcomes for the non-treated.

10.3.2 Difference in Difference Estimator in R

The difference in difference estimator is used to separate out the effect of time (v) from the treatment (x). This function creates a table with the two differences.

```
> f_did <- function(y,treat) {
+    y1 <- y[1,]
+    y2 <- y[2,]
+    did <- matrix(NA,3,3)
+    # creates the difference in difference matrix
+    did[1,1] <- mean(y2[treat==1])
+    # calculates the average outcomes for each
+    did[2,1] <- mean(y1[treat==1])
+    did[1,2] <- mean(y2[treat==0])
+    did[2,2] <- mean(y1[treat==0])
+    did[3,1] <- did[1,1] - did[2,1]
+    # calculates the differences.
+    did[3,2] <- did[1,2] - did[2,2]
+    did[1,3] <- did[1,1] - did[1,2]
+    did[2,3] <- did[2,1] - did[2,2]
+    did[3,3] <- did[3,1] - did[3,2]
+    row.names(did) <- c("Period 2", "Period 1", "Diff")
+    colnames(did) <- c("Treated", "Not Treated", "Diff")
+    return(did)
+ }
```

In the simulated data there are two groups, a treated group and a non-treated group. For the treated group, the outcome, y, is determined by both x and u. For the non-treated group the y is only determined by u. Of course both groups are affected by some random term e.

```
> set.seed(123456789)
> treat <- runif(N) < 0.5
> x2 <- x1 + treat
> # this time the change is 1 for the treated group.
> x <- rbind(x1,x2)
> y <- a + b*x + d*u + e
> did1 <- f_did(y,treat)
```

	Treated	Not Treated	Diff
Period 2	-22.16	-23.66	1.50
Period 1	-21.25	-19.77	-1.49
Diff	-0.90	-3.89	2.99

TABLE 10.2
Difference in difference estimator on simulated data.

The difference in difference estimator gives an estimate of 2.99, where the true value is 3. Table 10.2 presents the estimator, where the four cells in the top left are the mean of y for each case. Can you work out the bootstrapped standard errors for the estimator?

10.4   Minimum Wage Increase in New Jersey

What happens if the minimum wage is increased? Economic theory gives a standard prediction. If the minimum wage is increased above the equilibrium level wage, the demand for workers will fall and the supply of workers will increase. In fact, a minimum wage increase may actually harm more low-wage workers than it helps. While per-hour pay may increase, the number of hours may decrease. At least in theory.

What actually happens if the minimum wage is increased? The section uses the difference in difference estimator with data from Card and Krueger (1994) to determine what actually happened in New Jersey.

10.4.1   Data from Card and Krueger (1994)

Card and Krueger (1994) surveyed restaurants before and after the state of

New Jersey increased the minimum wage. In 1991, the US federal minimum wage was \$4.25/hour. In April of 1992, the state of New Jersey increased its minimum wage above the federal level to \$5.05/hour. In order to see how much a minimum wage increase led to a decrease in labor demand, Card and Krueger (1994) surveyed restaurants in New Jersey before and after the change. The following code imports the Card and Krueger (1994) data. Note that there is a labeling issue with two restaurants receiving the same label.

```
> x <- read.csv("cardkrueger.csv", as.is = TRUE)
> x1 <- cbind(x$SHEET,x$EMPFT,x$EMPPT,x$STATE,1)
> # SHEET is the firm ID.
> # this includes the initial employment levels.
> # column of 1s added to represent the initial time period.
> colnames(x1) <- c("SHEET","FT","PT","STATE","TIME")
> # FT - fulltime, PT - parttime.
> x1 <- as.data.frame(x1)
> x1$SHEET <- as.numeric(as.character(x1$SHEET))
> x1[x1$SHEET==407,]$SHEET <- c(4071,4072)
> # there is an issue with the labeling in the data.
> # two firms with the same label.
> x2 <- cbind(x$SHEET,x$EMPFT2,x$EMPPT2,x$STATE,2)
> # the second period of data.
> colnames(x2) <- c("SHEET","FT","PT","STATE","TIME")
> x2 <- as.data.frame(x2)
> x2$SHEET <- as.numeric(as.character(x2$SHEET))
> # a number of variables are changed into ``factors''
> # as.numeric(as.character()) changes them back into numbers.
> x2[x2$SHEET==407,]$SHEET <- c(4071,4072)
> x3 <- rbind(x1,x2)  # putting both periods together.
> colnames(x3) <- c("SHEET","FT","PT","STATE","TIME")
> x3$FT <- as.numeric(as.character(x3$FT))
> x3$PT <- as.numeric(as.character(x3$PT))
```

Figure 10.2 shows the minimum wage change will have some bite. A fairly large number of firms pay exactly the minimum of \$4.25 per hour. Most of the firms pay less than the proposed minimum of \$5.05.

10.4.2   Difference in Difference Estimates

The concern with just comparing employment in New Jersey before and after the minimum wage change is that other factors may have also changed between the two time periods. Card and Krueger (1994) use difference in difference to account for the time effects. The authors propose using restaurants in the neighboring state of Pennsylvania as the non-treated group. They argue that restaurants in Pennsylvania are not impacted by the New Jersey law change. In addition, these restaurants are similar enough to the New Jersey restaurants

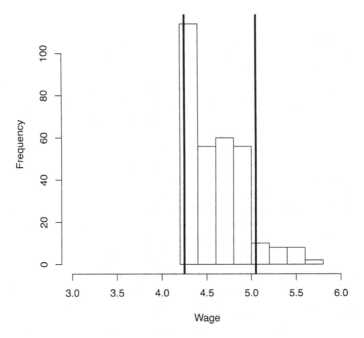

FIGURE 10.2
Histogram of New Jersey wage rate prior to the minimum wage increase. Note
that the minimum wage changes from $4.25 to $5.05 (represented by the two
vertical lines).

that other changes in the economy will be the same between the two states.
The restaurants in New Jersey and Pennsylvania will have the same "time
effect" on average.

To see what happens we can follow the procedure presented above.

```
> y <- rbind(x3$FT[x3$TIME==1] + x3$PT[x3$TIME==1],
+              x3$FT[x3$TIME==2] + x3$PT[x3$TIME==2])
> index_na <- is.na(colSums(y))==0
> y1 <- y[,index_na]
> treat <- x3$STATE[x3$TIME==1]==1
> treat1 <- treat[index_na]
> did2 <- f_did(y1,treat1)
```

The results suggest that the minimum wage increase has no impact on
employment in New Jersey. Table 10.3 presents the difference in difference
estimates on the total count of employees, both full time and part time.[2] In

---

[2]Card and Krueger (1994) present results on a measure called full-time equivalence. It is
unclear how that measure is calculated.

	Treated	Not Treated	Diff
Period 2	26.60	27.47	-0.87
Period 1	26.40	29.71	-3.31
Diff	0.20	-2.24	2.44

TABLE 10.3
Difference in Difference estimates of the effect of minimum wage law changes in New Jersey.

the table we actually see a very small increase in employment before and after the law change. Meanwhile, in Pennsylvania there is a slight decrease in employment over the same time period. The net results in a slight increase in employment associated with the change in the minimum wage law.

The result seems counter-intuitive, at least counter to standard economic theory. The result is heavily reliant on the assumption that the change in employment in Pennsylvania is a good proxy for the time effect in New Jersey. Is that assumption reasonable?

10.5 Fixed Effects

In data sets with more time periods we can use the fixed effects model. As above, it is assumed the time effect is the same for each individual, irrespective of the treatment. Some subset of the sample is **treated** and they receive the treatment in some subset of the time periods. Usually, there is a **pre-treatment** and a **post-treatment** period.

The section presents the general model with individual and time fixed effects.

10.5.1 Fixed Effects Estimator

As with the difference in difference, the time effect is assumed to be the same for everyone. This model allows individuals to have different outcomes that are persistent through time. In the restaurant example, we expect some restaurants to have higher employment in both periods, relative to other restaurants. However, with only two time periods we cannot account for these differences. With more pre-treatment time periods we can attempt to measure differences between restaurants. Accounting for such differences may be particularly important if Pennsylvania and New Jersey tend to have different types of restaurants.[3]

The general model has the observed outcome as a function of two fixed effects. The first is an individual effect that is allowed to vary across individuals but is persistent through time. The second is a time effect that varies across time but is persistent through the cross section of individuals. There is also a treatment effect that is assumed to affect a subset of individuals in a subset of time periods and an unobserved characteristic that varies across individuals and time.

$$y_{it}(x_{it}) = \alpha_i + \beta x_{it} + \gamma_t + \epsilon_{it} \tag{10.7}$$

where

$$x_{it} = \begin{cases} 1 \text{ if } i \text{ is treated and } t > T_0 \\ 0 \text{ otherwise.} \end{cases} \tag{10.8}$$

where the treatment occurs in period T_0, y_{it} is the outcome, α_i is the fixed effect for individual i, β measures the treatment effect, γ_t is the "time effect," x_{it} is the indicator of the treatment and ϵ_{it} is the unobserved term for each individual and time period. We are interested in estimating β.

Consider our restaurant example but this time we observe employment levels for the restaurants for a number of years prior to the minimum wage increase in New Jersey. We can allow different restaurants to be big or small leading to a different number of employees on average. We can also allow all restaurants to be hit by changes to general economic conditions. Lastly, we assume that only restaurants in New Jersey are impacted by the increase in the minimum wage in New Jersey.

10.5.2 Nuisance Parameter

A nice feature of the fixed effects model is that it accounts for differences between the treated and non-treated groups that are not due to the treatment itself. The α_i parameter is allowed to be different for each individual i. While allowing a lot of flexibility this parameter is also a nuisance. It is a nuisance parameter in the econometrics sense. It cannot be **consistently** estimated. It does not converge to the true value as the number of individuals gets large.[4] In addition, it may not be of direct relevance. As mentioned above, we are interested in estimating β not α_i. It is also a nuisance parameter in that it makes the model difficult to estimate.

A simple solution is to not include the individual dummy in the estimator. The problem with this approach is that it makes the estimator much noisier and less likely to be correct. This is because all the individual dummy variables have been added to the error term. It may also lead to a biased estimate if there

---

[3]Chapter 12 uses a mixture model to account for this heterogeneity.

[4]It can be consistently estimated as the number of time periods gets large, but microeconometric panels tend not to have a large number of time periods. See Appendix A for a discussion of this property.

is some systematic difference between the treated group and the non-treated group.

10.5.3 Adjusted Fixed Effects Estimator

One nice solution is to do adjusted estimation. In the first step, the pre-treatment data is used to estimate the α_i parameter. In the second step, the estimates from the first step are used to estimate the treatment effect parameter of interest. The pre-treatment outcomes are averaged.

$$\frac{1}{T_0} \sum_{t=1}^{T_0} y_{it} = \alpha_i + \frac{1}{T_0} \sum_{t=1}^{T_0} \gamma_t + \frac{1}{T_0} \sum_{t=1}^{T_0} \epsilon_{it} \tag{10.9}$$

If we substitute that estimate back into Equation (10.7) we get the following equation. Note that the nuisance parameter has been removed (it has been replaced with an estimate).

$$y_{it} - \bar{y}_i = \beta x_{it} + \gamma_t - \bar{\gamma} + \epsilon_{it} - \bar{\epsilon}_i \tag{10.10}$$

where $\bar{y}_i$ is the average outcome for individual i over the pre-treatment period. Equation (10.10) gives the basis for the adjusted fixed effects estimator.

A potential issue is that we have added extra noise to the estimator ($\bar{\epsilon}_i$). Under standard assumptions, the larger the pre-treatment period the smaller this problem is.[5]

10.5.4 Two Step Fixed Effects Estimator

Another approach is a two step estimator. In the first step, the individual fixed effects are estimated for each individual separately in the pre-treatment period. The individual fixed effects estimator is simply the mean of the residual of the regression of the outcome variable on the time-dummies.

$$\hat{\alpha}_i = \frac{1}{T_0} \sum_{t=1}^{T_0} (y_{it} - \frac{1}{N} \sum_{j=1}^{N} y_{jt}) \tag{10.11}$$

The second step regresses the outcome less the fixed effect on the policy variables of interest.

$$y_{it} - \hat{\alpha}_i = \gamma_t + \beta x_{it} + (\alpha_i - \hat{\alpha}_i) + \epsilon_{it} \tag{10.12}$$

Equation (10.12) forms the basis for the estimator with the outcome and the "error" term adjusted by netting out the estimated fixed effect.

This estimator highlights the estimation problem. If the pre-treatment period is not very long then each of the $\hat{\alpha}_i$ will not be well estimated. Under

---

[5]The average $\bar{\epsilon}_i$ will tend to zero. See Chapter 1 and Appendix A.

standard assumptions these estimates are unbiased, which implies that in some cases the inaccuracy of the estimates are fine, while in other cases it may be problematic.

Also note the similarity and difference between the two step estimator and the adjusted estimator.[6]

10.5.5   Fixed Effects Estimator in R

The following fixed effects estimator takes data in the form of panel data matrices and converts to use `lm()`. The function takes advantage of **R**'s ability to quickly create a large number of dummy variables using `as.factor()`. Note the function has the option to create fixed effects *only* in the time dimension. It also does not need to include a treatment variable.

```
> f_fe <- function(Y, X=NULL, cross=TRUE) {
+    Y <- as.matrix(Y)
+    T <- dim(Y)[1]
+    N <- dim(Y)[2]
+    XT <- matrix(rep(c(1:T),N), nrow=T)
+    # creates a T x N matrix with numbers 1 to T in each row
+    y <- as.vector(Y)
+    t <- as.vector(XT)
+    # set up for different cases
+    if (cross) { # create cross-section dummies
+      XC <- t(matrix(rep(c(1:N),T), nrow=N))
+      # creates a T x N matrix with 1 to N in each column
+      c <- as.vector(XC)
+    }
+    if (is.null(X)==0) { # create treatment variable
+      X <- as.matrix(X)
+      treat <- as.vector(X)
+    }
+    # estimator
+    if (cross & is.null(X)==0) { # standard case
+      lm1 <- lm(y ~ treat + as.factor(t) + as.factor(c))
+    }
+    else {
+      if (is.null(X)==0) { # no cross-section
+        lm1 <- lm(y ~ treat + as.factor(t))
+      }
+      else { # no treatment
+        lm1 <- lm(y ~ as.factor(t))
```

---

[6]An alternative approach to dealing with the nuisance estimator is to assume that the nuisance parameter has a particular distribution. This approach is generally called "random effects."

```
+     }
+   }
+   return(lm1)
+ }
```

The simulated panel data set has 100 individuals observed over 10 time periods. About half of the individuals are treated and the treatment occurs in the last time period.

```
> set.seed(123456789)
> N <- 100
> T <- 10
> alpha <- runif(N)
> gamma <- runif(T)
> beta <- 3
> epsilon <- matrix(rnorm(N*T),nrow=T)
> treat <- runif(N) < 0.5
> y <- t(matrix(rep(alpha,T),nrow = N)) + gamma + epsilon
> y[1,] <- y[1,] + beta*treat
> treat1 <- matrix(0,T,N)
> treat1[1,] <- treat
```

We can compare the different estimators on the simulated data.

```
> # standard estimator
> lm1 <- f_fe(y, treat1)
> # No individual fixed effects estimator
> lm2 <- f_fe(y, treat1, cross = FALSE)
> # Adjusted estimator
> y0 <- y[2:T,]  # pre-treatment outcomes.
> alpha_hat <- colMeans(y0) # calculate alpha.
> y2 <- y - t(matrix(rep(alpha_hat,T),nrow=N))
> # adjust outcome.
> lm3 <- f_fe(y2, treat1, cross = FALSE)
> # Two step estimator
> lm4 <- f_fe(y0, cross = FALSE)  # adjust for time effects.
> y0_res <- matrix(lm4$residuals, nrow=T-1)
> alpha_hat <- colMeans(y0_res) # calculate alpha.
> y3 <- y - t(matrix(rep(alpha_hat,T),nrow=N)) # adjust outcome
> lm5 <- f_fe(y3, treat1, cross=FALSE)
```

Table 10.4 provides a nice comparison between the four approaches. The true parameter is 3 and the standard model gives a good estimate. The model in which the individual dummies are dropped gives a relatively poor estimate. The other models give a good estimate.

```
> require(stargazer)
> stargazer(list(lm1,lm2,lm3,lm5),keep.stat = c("n","rsq"),
+          keep=c("treat"), float = FALSE)
```

		Dependent variable:		
		y		
	(1)	(2)	(3)	(4)
treat	2.998***	3.060***	2.998***	2.998***
	(0.215)	(0.214)	(0.195)	(0.195)
Observations	1,000	1,000	1,000	1,000
R^2	0.458	0.340	0.378	0.378
Note:			*p<0.1; **p<0.05; ***p<0.01	

TABLE 10.4

Results from the various fixed effects estimators on simulated data where the true value is 3. The models are the standard estimator (1), the estimator where the individual dummies are dropped (2), the adjusted estimator (3), and the two-step estimator (4).

10.6   Effect of a Federal Minimum Wage Increase

In 2007, the George W. Bush administration increased the federal minimum wage to $7.25. The results from the difference in difference analysis of Card and Krueger (1994) suggest that this change will have little or no effect on employment. However, the approach makes quite strong assumptions and looks at the impact on restaurants rather than on individuals.

The NLSY97 data seems well suited to analyzing the impact of Bush's federal minimum wage increase. The individuals in the data are in their late 20s and early 30s when the changes occur. At least some proportion of these individuals are likely to work in jobs that pay minimum wage or would be impacted by the changes. The analysis here follows Canadian-American labor economist Janet Currie, who uses NLSY79 to analyze the impact of minimum wage changes in 1979 and 1980 (Currie and Fallick, 1996).

10.6.1 NLSY97

The following code imports a data set that I have created from NLSY97.[7] It then creates two matrix panels, one for income and one for hours worked. For each, the rows are time periods and the columns are individuals.

```
> x <- read.csv("NLSY97Panel.csv",as.is=TRUE)
> x_names <- read.csv("NLSY97Panel_names.csv",
+                     header=FALSE,as.is = TRUE)
> colnames(x) <- as.vector(x_names[,1])
> # create two matrices, with 18 (years)
> # rows 8984 (individuals) columns
> # one for income and one for hours worked.
> year <- c(1997:2014)
> year1 <- c("97","98","99","00","01","02","03",
+            "04","05","06","07","08",
+            "09","10","11","12","13","14")
> # below we need both versions of year.
> W <- Y <- matrix(NA,18,8984)
> for (i in 1:18) {
+    hrs_name <- paste("CVC_HOURS_WK_YR_ALL_",
+                      year1[i],"_XRND",sep="")
+    # paste() is used to concatenate strings.
+    inc_name <- paste("YINC_1700_",year[i],sep="")
+    if (hrs_name %in% colnames(x)) {
+      Y[i,] <- ifelse(x[,colnames(x)==hrs_name] >= 0,
+                      x[,colnames(x)==hrs_name],NA)
+    }
+    # %in% asks whether something is an element of a set.
+    if (inc_name %in% colnames(x)) {
+      W[i,] <- ifelse(x[,colnames(x)==inc_name] >= 0,
+                      x[,colnames(x)==inc_name],NA)
+    }
+ }
> rate_07 <- W[11,]/Y[11,]
> # calculates the wage rate for each person.
> x$treat <- ifelse(rate_07<7.26 | W[11,]==0 | Y[11,]==0,1,0)
> # treated if earn less than 7.26/hour or no earnings in 2007.
```

The data includes information on almost 9,000 individuals who are tracked across 18 years. For each individual we know their income and hours worked for the year. From this we can calculate their average hourly wage rate. We can also determine whether there were some individuals in the data earning less than $7.26 an hour in 2007. We will call this group the **treated group**. Individuals

---

[7]The data is available here: https://sites.google.com/view/microeconometricswithr/table-of-contents.

earning more than this are assumed to be unaffected by the policy change. These are the **non-treated group**. How reasonable is this assignment? Can you re-do this analysis with a different assignment to treated and non-treated?

10.6.2 Fixed Effects Estimators of the Minimum Wage

We can use the standard fixed effects estimator to determine the effect of the minimum wage change on hours worked. The code creates a treatment variable. Note that the pre-treatment period is the first ten years, while the post-treatment period is the last 8 years.

```
> N <- 600 # reduce the size for computational reasons.
> T <- 18
> y1 <- Y[,1:N]
> treat1 <- matrix(0,T,N)
> for (i in 11:T) {
+     treat1[i,] <- x$treat[1:N]
+ }
> lm1 <- f_fe(y1, treat1)
```

As discussed above, the individual dummy variables are a nuisance to estimate. You can see this by changing the number of individuals used in the estimation. As the number increases the computation takes longer and longer.

It is lot less computationally burdensome to estimate the "adjusted" fixed effects estimator. The outcome is "adjusted" by differencing out the average hours worked for each individual in the pre-treatment period.

```
> N <- 8984  # full data set
> y0 <- Y[1:10,1:N]  # pre-treatment
> alpha_hat <- colMeans(y0, na.rm = TRUE)
> y1 <- Y - t(matrix(rep(alpha_hat,T), nrow = N))
> treat1 <- matrix(0,T,N)
> for (i in 11:T) {
+     treat1[i,] <- x$treat[1:N]
+ }
> lm2 <- f_fe(y1, treat1, cross=FALSE)
```

We can also estimate the two step fixed effects estimator in order to compare the results.

```
> lm3 <- f_fe(y0, cross=FALSE)
> y0_res <- y0 - matrix(rep(lm3$coefficients,N),nrow=10)
> alpha_hat <- colMeans(y0_res, na.rm = TRUE) # calculate alpha.
> y3 <- Y - t(matrix(rep(alpha_hat,T),nrow=N)) # adjust outcome
> lm4 <- f_fe(y3, treat1, cross=FALSE)
```

We can compare results across the three estimators. The estimates are similar. They show a fairly substantial reduction in hours associated with the increase in the minimum wage.

	Dependent variable:		
	y		
	(1)	(2)	(3)
treat	−296.386***	−271.147***	−274.277***
	(40.853)	(7.821)	(7.797)
Observations	8,297	127,107	127,107
R^2	0.542	0.275	0.283
Note:			*p<0.1; **p<0.05; ***p<0.01

TABLE 10.5
Fixed effects estimate of the effect of increasing the minimum wage on hours worked. The models are the full fixed effects model with a subset of 600 individuals (1), the full data set but adjusted to average out the individual fixed effects (2), and the two-step estimator (3).

Table 10.5 presents the estimates with the standard estimator and a subset of the individuals, as well as the adjusted estimators with all individuals. It shows a 10% reduction in hours for a full-time person working 2,000 hours, but much more for an average individual in the sample.

10.6.3 Are Workers Better Off?

While the decrease in hours is large, it is not clear that these workers are worse off because the wage increase was substantial.

```
> a <- mean(x[x$treat==1,]$YINC_1700_2007, na.rm = TRUE)
> b <- mean(x[x$treat==1,]$CVC_HOURS_WK_YR_ALL_07_XRND,
+          na.rm = TRUE)
> c <- a/b   # 2007 wage rate
> d <- mean(x[x$treat==1,]$YINC_1700_2010, na.rm = TRUE)
> # 2010 income
> e <- mean(x[x$treat==1,]$CVC_HOURS_WK_YR_ALL_10_XRND,
+          na.rm = TRUE)
> # 2010 hours
> d - c*(e + 270)

[1] 4206.07
```

```
> # actual less counter-factual
```

The minimum wage increase leads to lower hours but higher incomes. To see what happens to the average treated person, compare their actual income in 2010 to their estimated counterfactual income in 2010. Their actual average income in 2010 is $12,548. Their actual hours worked is 1034. According to our estimates if the minimum wage had not increased, they would have worked 1304 hours. However, they would have earned less. On average their wage prior to the minimum wage increase was $6.40 per hour. Assuming that is the counter-factual wage, then the minimum wage change increased their income by $4,200 per year. According to this analysis, the policy increased income for the treated group despite lowering their hours worked. What criticisms would you have of this analysis?

10.7 Discussion and Further Reading

The last part of the book presents the third major approach to estimating the policy variables of interest, repeated measurement. This chapter considers panel data. In this data, we observe outcomes for a large number of the same individuals over a number of time periods.

This data allows us to estimate treatment effects by comparing outcomes before and after the treatment. The problem with this comparison is that time may have caused other changes to occur. The observed difference is affected by both the treatment and time. The panel structure suggests a solution. If there are some individuals who were unaffected by the treatment then only time affected the difference in outcomes. We can use these non-treated individuals to measure the impact of time and difference it out. The classic paper is Card and Krueger (1994).

If we have enough time periods we can account for heterogeneity in the observed outcomes. We can use **fixed effects** to account for unobserved differences in outcomes between individuals. The next chapter considers panel data with some of the assumptions relaxed. In particular, it considers the synthetic control approach of Abadie et al. (2010). Chapter 12 accounts for heterogeneity with a mixture model. The chapter revisits Card and Krueger (1994).

11

Synthetic Controls

11.1 Introduction

By observing the same individual with different levels of exposure to a policy we can learn about how the policy affects that individual. However, we have a problem that "time" is also affecting the individual. Chapter 10 showed how panel data could be used to "control for time" using **fixed effects**.

In the standard fixed effects model, it is assumed that time has the same effect on all individuals. In particular, time has the same effect for both **treated** and **non-treated groups** (or **control group**). This assumption is sometimes called **parallel trends**. Abadie et al. (2010) argue that this assumption can be weakened. The authors argue for choosing the **control group** more carefully. The authors suggest using the **pre-treatment period** to estimate the relationship between members of the **treated group** and members of the **control group**. Instead of placing equal weights on every member of the control group, members of the treated group should place more weight on members of the control group that are similar.

This chapter introduces the synthetic control estimator. It presents the general model, discusses the identification problem and presents the solution provided by Abadie et al. (2010). The chapter presents two alternative approaches, LASSO and factor models. The estimators are used to determine the effect of the federal minimum wage increase discussed above.

11.2 Beyond "Parallel Trends"

The fixed effects model presented in Chapter 10 makes the very strong **parallel trends** assumption. The impact of time is *identical* for everyone. This does not seem reasonable. We would like to estimate a model that allows the impact of time to vary, in particular to vary across treated and non-treated groups.

11.2.1   A General Fixed Effects Model

One idea is to allow the fixed effects model to be more general. Below we make two changes to the model.

$$y_{it}(x_{it}) = \alpha_i + \beta_{it}x_{it} + \gamma_t + \delta_{it} + \epsilon_{it} \qquad (11.1)$$

where, as before, α_i is the individual fixed effect, γ_t is the time-effect and β_{it} is the treatment effect. The difference now is that there is an additional time effect, δ_{it} that varies over time and across individuals. Similarly, β_{it} measures the treatment effect and may vary across individuals and over time.

These two relatively simple changes to the model have a fundamental impact on our ability to estimate the treatment effect. Under this model, the individual treatment effect estimation is given by the following equation.

$$y_{i1}(1) - y_{i1}(0) = \beta_{i1} + \delta_{i1} - \delta_{i0} + \epsilon_{i1} \qquad (11.2)$$

Note that we simplified things and assumed we can accurately estimate the individual and time fixed effects. The value of interest, β_{i1} is confounded by the individual time varying changes represented by $\delta_{i1} - \delta_{i0}$. Because this time varying effect is individual specific, there is no obvious method for estimating it.

11.2.2   A Slightly Less General Model

Unfortunately, the generalization of the fixed effects model made above means that we can no longer estimate the model. At least we cannot estimate the policy variable of interest. Consider the following slightly less general model.

$$y_{it}(x_{it}) = \beta_{it}x_{it} + f_{1t}\lambda_{1i} + f_{2t}\lambda_{2i} + \epsilon_{it} \qquad (11.3)$$

where f_{kt} represents some "factor" that determines the outcome of interest and λ_{ki} is the weight that individual i places on the factor. If we are considering hours worked by individual in NLSY97 then these factors may account for employment in the region where they live or the industry they work in. In Equation (11.3) there are two factors. These factors affect the hours worked for everyone in the data but each individual may be affected more or less by each factor. For example, if someone lives in Texas they may be more affected by a large fall in the price of petroleum than someone in Oregon.

11.2.3   Synthetic Synthetic Control Data

The synthetic data is generated according to a **factor model**.[1] All simulated individuals have their outcome determined by three factor values. However, individuals weight these factor values differently. There are 35 time periods and 200 individuals.

---

[1]In particular, it is generated using a **convex factor model**, which is discussed in more detail below.

```
> set.seed(123456789)
> N <- 200  # individuals
> T <- 35  # time periods
> K <- 3 # factors
> sigma <- 0.1  # noise
> F <- matrix(3*runif(T*K), nrow=T) # factors
> Lambda <- matrix(exp(rnorm(N*K)), nrow=K)
> # weights are positive
> Lambda <- t(t(Lambda)/rowSums(t(Lambda)))
> # sum weights to 1
> E <- matrix(rnorm(T*N,mean=0,sd=sigma),nrow=T) # unobserved
> Ybar <- F%*%Lambda  # average outcome
> Ys <- Ybar + E  # actual outcome
```

Note there is no policy effect in the simulated data. Each individual's outcome is determined by their individual weights on three factors, plus an unobserved term.

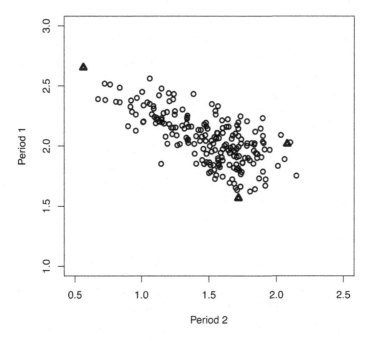

FIGURE 11.1
Plot of outcomes for periods 1 and 2. Triangles are located at the factor value vectors for the two periods.

Figure 11.1 presents a plot of outcomes for the 200 individuals in periods 1 and 2. Each individual in the data places a weight on the three factor values, which are represented by the triangles. The implication is that while each individual is unique there are other individuals in the data that have similar outcomes because they weight the factor values in the same way.

11.2.4 Constructing Synthetic Controls with OLS

The goal is to use the non-treated individuals to construct the "time effect" for the treated individuals. As in Chapter 10, once we have accounted for the time effect, we can back out the treatment effect. But how do we construct this **synthetic control**?

Perhaps the most straightforward way is to use OLS. We discussed in Chapter 1 that OLS captures the correlation between variables. Let's say we want to construct a "synthetic" version of individual 1 using data from individuals 2 to 5. We can think of the outcomes for individual 1 as the variable we want to explain, while the outcomes for individuals 2 to 5 are the explanatory variables.

$$y_{1t} = \omega_0 + \omega_2 y_{2t} + \omega_3 y_{3t} + \omega_4 y_{4t} + \omega_5 y_{5t} + \epsilon_{1t} \qquad (11.4)$$

Equation (11.4) assumes that the observed outcome for synthetic individual 1 in period t is a weighted average of the outcomes for the other four individuals in the same time period.

We can then write this out in matrix form.

$$\mathbf{y}_1 = \mathbf{Y}_{-1}\omega + \epsilon \qquad (11.5)$$

where $\mathbf{y}_1$ is the vector of outcomes for synthetic individual 1 across all the time periods, $\mathbf{Y}_{-1}$ is the matrix of outcomes for all the individuals *not* individual 1 across all the time periods, finally ω is the vector of weights that individual 1 places on the other synthetic individuals.

Following the same matrix algebra we used in Chapters 1 and 2, OLS determines the weights we should use for our synthetic control.

$$\hat{\omega} = (\mathbf{Y}'_{-1}\mathbf{Y}_{-1})^{-1}\mathbf{Y}'_{-1}\mathbf{y}_1 \qquad (11.6)$$

With these estimated weights the synthetic individual 1's outcome in some future period is a weighted average of the outcomes for the other synthetic individuals.

$$\hat{y}_{1(t+s)} = \mathbf{Y}'_{-1(t+s)}\hat{\omega} \qquad (11.7)$$

11.2.5 OLS Weights in R

We can use the first 30 time periods in the synthetic data constructed above to estimate the synthetic control for individual 1. We can use the last 5 time periods to compare our synthetic prediction to the actual outcomes.

```
> lm1 <- lm(Ys[1:30,1] ~ Ys[1:30,2:5])
> omega_hat <- lm1$coefficients
> y1_hat_ols <- cbind(1,Ys[31:35,2:5])%*%omega_hat
```

Figure 11.6 plots the actual outcomes and predicted outcomes for synthetic individual 1 in periods 31 to 35. The 45 degree line represents perfect prediction. The chart shows that our OLS based synthetic control tracks the actual outcomes pretty well.

The issue with this method is that our choice of using the next four individuals is completely arbitrary. We can construct more accurate controls using more individuals.

11.2.6   A "Wide" Data Problem

To understand the issue consider what happens when we use the next 28 simulated individuals. The loop finds the **root mean squared error** as each individual is added to the regression.

```
> tab_res <- matrix(NA,28,2)
> for (i in 2:29) {
+     lm1 <- lm(Ys[1:30,1] ~ Ys[1:30,2:i])
+     omega_hat <- lm1$coefficients
+     res <- Ys[31:35,1] - cbind(1,Ys[31:35,2:i])%*%omega_hat
+     tab_res[i-1,1] <- i
+     tab_res[i-1,2] <- mean(res^2)^(0.5)
+ }
> # creates a table of the rmse for different numbers of
> # explanatory variables.
```

Figure 11.2 presents a measure of predictive ability of the synthetic control. The chart uses root mean squared error of the difference between the actual outcome and the predicted outcome. The chart suggests that the prediction is optimized when 5 to 10 other individuals are used. The prediction gets slowly worse as more individuals are used. Why would that be? Then around 25, the prediction starts to go off the rails. Why is that?

We are running into a problem brought up in Chapter 2. We have collinearity and multicollinearity problems. Try running the regression hashed out below. It should produce some sort of error. Probably something related to not being able to find the inverse of the matrix. There is a collinearity problem when the number of "explanatory variables" (simulated individual outcomes) equals the number of "observations" (time periods). As we add in more individuals the correlation between individuals in the data increases until the point where the regression starts producing garbage and then it finally stops working all together.

```
> y1 <- Ys[1:30,1]
```

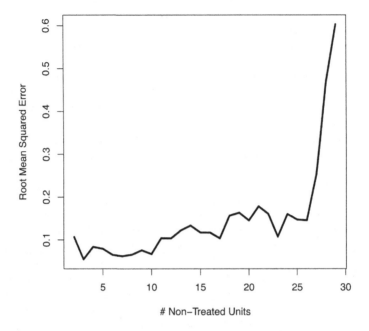

FIGURE 11.2
Plot of root mean squared error for randomly choosing from the non-treated individuals.

```
> Y_n1 <- Ys[1:30,-1]
> #omega_hat <- solve(t(Y_n1)%*%Y_n1)%*%t(Y_n1)%*%y1
```

The problem of how to choose between potentially collinear explanatory variables is called a **wide data problem**. Here "wide" refers to the fact that our data matrix is very wide relative to its length (the number of observations or time periods). In our simulated data we have 199 explanatory variables and 30 observations. The rest of the chapter discusses three solutions to the wide data problem; the Abadie et al. (2010) estimator, the LASSO and convex factor models.

11.3   Abadie Estimator

Abadie et al. (2010) provide one popular solution to the wide data problem presented above. Their idea is to restrict the weights. In particular, the authors restrict the weights to be positive and sum to one. The treated unit is assumed to be a convex combination of the non-treated units.

11.3.1 Restricting Weights

Abadie et al. (2010) present an OLS estimator like the one presented in the previous section. However, the weights are restricted such that each treated individual is a convex combination of the non-treated individuals. We assume the outcomes of individual 1 are a convex combination of the outcomes for the other 199 simulated individuals.

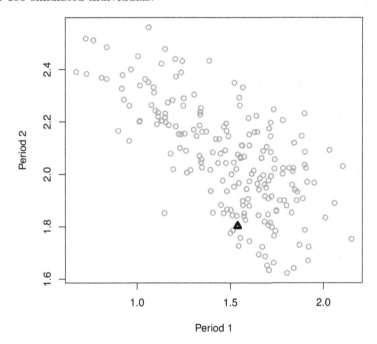

FIGURE 11.3
Plot of simulated outcomes for period 1 and period 2. The triangle is individual 1.

Figure 11.3 shows that individual 1 sits in the cloud of simulated outcomes for period 1 and period 2. This means that it is possible to represent individual 1's outcomes as a convex combination of the other simulated individuals.

The authors assume that $\omega_0 = 0$, $\omega_j \geq 0$ and $\sum_{j=2}^{N} \omega_j = 1$. That is, the constant term in the OLS is zero and the remaining weights are positive and add to 1.

The Abadie et al. (2010) estimator solves the following constrained least squares problem.

$$\min_{\hat{\omega}} \quad \sum_{t=1}^{T_0} \left(y_{1t} - \sum_{j=2}^{N} \hat{\omega}_{1j} y_{jt}\right)^2$$

$$s.t. \quad \sum_{j=2}^{N} \hat{\omega}_j = 1 \tag{11.8}$$

$$\hat{\omega}_j \geq 0 \text{ for all } j \in \{2, ..., N\}$$

The authors suggest that the restriction can be justified by assuming a factor model is generating the observed data. Note however, there is no relationship between the assumptions presented here and the assumptions on the factor model presented below.

11.3.2   Synthetic Control Estimator in R

Equation (11.8) provides the pseudo-code for the estimator in **R**.

```
> f_sc <- function(par,yt,ynt) {
+    yt <- as.matrix(yt)
+    ynt <- as.matrix(ynt)
+    omega <- exp(par)/(sum(exp(par)))
+    # the restriction on the weights.
+    sos <- mean((yt - ynt%*%omega)^2)
+    # sum of squares
+    return(sos)
+ }
> J <- dim(Y_n1)[2]
> a <- optim(par=rnorm(J), fn=f_sc, yt=y1, ynt=Y_n1,
+               control = list(trace=0,maxit=1000000))
> omega_hat <- exp(a$par)/sum(exp(a$par))
> y1_hat_sc <- Ys[31:35,-1]%*%omega_hat
```

Figure 11.6 shows that the Abadie et al. (2010) estimator is able to predict pretty well. Like the OLS estimator presented above, the prediction/actual pairs lie along the 45 degree line. The big difference is that the Abadie et al. (2010) estimator does not need to be restricted, at least not in this case.[2]

11.4   Regularization

The wide data problem is difficult for standard estimators, but it is the type of problem machine learning estimators excel at. This section presents a popular

---

[2]Eventually, the estimator also runs into collinearity and multicollinearity problems Doudchenko and Imbens (2016).

machine learning estimator called the **LASSO**. This is a **regularized estimator**. This means that some "cost" is added to the least squares problem. There is an explicit cost or penalty to adding an additional explanatory variable.

11.4.1 Turducken Estimation

A turducken is a dish in which a chicken is stuffed inside of a duck, which is then stuffed inside a turkey. Many machine learning algorithms have this property of estimation within an estimation. The "chicken" layer is the standard estimation problem in which the **tuning parameters** are taken as given and not estimated. The chicken layer uses some standard criterion such as **least squares** or **maximum likelihood**. The "duck" layer takes the results from the chicken layer and determines the appropriate tuning parameter values. The duck layer uses a measure like **root mean squared error** of the model on the test data to evaluate the tuning parameters. There may even be a turkey layer in which different machine learning models are evaluated. This layer may add a criterion such as speed of estimation or accuracy on new test data withheld from the original problem. Here we limit things to "ducken" estimation.

11.4.2 LASSO

The Stanford statistician, Robert Tibshirani, introduced the idea of a **regularized** OLS estimator he called LASSO for **least absolute shrinkage and selection operator** (Tibshirani, 1996). In the LASSO, the parameters are regularized through a penalty term which is the sum of their absolute values.[3] The problem involves solving a constrained optimization problem.

The chicken layer is a standard least squares estimator. Note that the chicken layer is estimated on a subset of time periods $(T - s)$. This is the **training data**. The remaining s time periods are **held-out** and not used in the estimation. The held-out data is the **testing data**.

$$\hat{\omega}_\theta = \arg\min_\omega \quad \sum_{t=1}^{T-s}(y_{1t} - \omega_0 - \sum_{j=2}^N \omega_j y_{jt})^2 + \theta \sum_{j=-1} |\omega_j| \quad (11.9)$$

where θ is described as a **tuning parameter**. The higher the value of θ the more pressure to reduce the sum of the absolute values of the parameters. This estimator will tend to set "unimportant" weights to zero. It tends to set non-identified weights to zero and also set small but identified weights to zero.[4]

The duck layer uses the **held-out sample** and chooses θ to minimize **root mean squared error**.

$$\min_{\hat{\theta}} \quad \left(\tfrac{1}{s}(\sum_{t=T-s+1}^{T_0}(y_{1t} - \hat{\omega}_{0\hat{\theta}} - \sum_{j=2}^N \hat{\omega}_{j\hat{\theta}} y_{jt})^2)\right)^{0.5} \quad (11.10)$$

---

[3] Another standard regularization procedure is to have a penalty equal to the square of the values. This is called a ridge regression.

[4] From Chapter 1 we know that the standard OLS model gives unbiased estimates. If the model presented above is true, then this regularized OLS model must give biased estimates of the true weights. The importance of this is up for debate.

where ω_θ is the vector of weights that is the solution to Equation (11.9).

One argument for using root mean squared error as the appropriate criteria is that it provides a trade-off between two things we may care about in our prediction. It weights the bias, that is the extent our estimate is wrong in expectation, and the variance, that is the extent that our estimate is wrong in fact.

11.4.3 LASSO in R

Using Equation (11.9) as pseudo-code we have a simple regularized OLS estimator. The estimator defaults to $\theta = 0$, but this is changed in the procedure below. Note that code uses the matrix notation because it is a bit more compact.

```
> f_ols_reg <-function(omega,y_t,Y_nt,th1 = 0) {
+    y_t <- as.matrix(y_t)
+    Y_nt <- as.matrix(Y_nt)
+    sos <- (y_t - cbind(1,Y_nt)%*%omega)^2 + th1*sum(abs(omega))
+    return(mean(sos, na.rm = TRUE))
+ }
```

The code splits the data into a training data set, which is the first 25 periods, and a test data set which is the next 5 periods. We will compare predictions using the last 5 periods.

```
> y1_r <- y1[1:25]  # training data
> y1_t <- y1[26:30]  # test data
> Y_n1_r <- Y_n1[1:25,]
> Y_n1_t <- Y_n1[26:30,]
```

The following loop goes through the set of **tuning parameters** and finds the one that minimizes the root mean squared error.

```
> f_lasso <- function(y_tr,y_tt,Y_ntr,Y_ntt,K=3) {
+    y_tr <- as.matrix(y_tr)
+    y_tt <- as.matrix(y_tt)
+    Y_ntr <- as.matrix(Y_ntr)
+    Y_ntt <- as.matrix(Y_ntt)
+    set.seed(123456789)
+    theta_values <- c(1:K)-1
+    J <- dim(Y_ntr)[2]
+    omega_th <- NULL
+    rmse_old <- 10000  # some large number.
+    for (k in 1:K) {
+      ai <- optim(par=rep(0,J+1), fn=f_ols_reg, y_t=y_tr,
+               Y_nt=Y_ntr, th1 = theta_values[k],
+               control = list(maxit=10000))
```

```
+      omega <- ai$par
+      rmse_new <- mean((y_tt - cbind(1,Y_ntt)%*%omega)^2)^(0.5)
+      if (rmse_new < rmse_old) {
+        omega_th <- omega
+        rmse_old <- rmse_new
+        #print(rmse_new)
+      }
+    }
+    return(omega_th)
+ }
> omega_hat <- f_lasso(y1_r,y1_t,Y_n1_r,Y_n1_t)
> y1_hat_lasso <- cbind(1,Ys[31:35,-1])%*%omega_hat
```

Testing the LASSO on the synthetic data shows that the estimator does okay. Figure 11.6 shows that in the example it may have the best predictions of each of the models tested.

11.5 Factor Models

We started off the chapter by saying that we could make the fixed effects model more general by adding a factor model. We call this an **interactive fixed effects** model (Bai, 2009). However, we then presented three other methods for estimating the synthetic control weights. We seemed to forget that we had a factor model.

Why didn't we simply estimate the factor model? The answer is easy. It is hard.

As a reminder, the problem with estimating the synthetic control is that we don't have enough observations relative to the number of potential variables. This is called a **wide data problem**. Factor models have been suggested as a solution to this problem for close to 100 years.

Factors models date back to at least Hotelling (1933), although he preferred the term "component models." Hotelling didn't like the fact that he was using matrix factorization to factor a factor model, the result of which is that today, economists use "principal components to factor factor models." I guess we didn't read the memo the whole way through.

Hotelling's contribution was to formalize a method in use for solving a wide data problem in psychometrics. Researchers had individual subjects take a battery of tests, so they had data with a small number of individuals and a large number of test results.

Variations on the model presented below are popular in machine learning, under the names **topic models** or **non-negative matrix factorization**. In machine learning these models are categorized as **unsupervised learning models**, where LASSO is **supervised learning models** (Blei, 2012; Huang et al., 2014).

So why is it hard? Hotelling (1933) pointed out that the matrix factorization problem at the heart of the approach is not identified. There are many nontrivial ways to factor one matrix into two matrices. You may remember that you can factor the number 12 into a number of different factor pairs: 1 and 12, 2 and 6, and 3 and 4. It is similar for matrix factorization. There are many pairs of matrices that can be multiplied together to get the same observed matrix. In trying to solve one identification problem we have run headlong into a new (actually very old) identification problem.

Fu et al. (2016) suggests a solution to the new identification problem. The paper suggests restricting the factor weights to be positive and sum to 1. This is called a convex matrix factorization problem. Fu et al. (2016) and Adams (2018) suggest that this restriction significantly reduces the identification problem. Under certain restrictions, it may even provide a unique solution to the matrix factorization problem.

11.5.1 Matrix Factorization

Consider we observe some matrix $\mathbf{Y}$ that is $T \times N$. This may represent a panel data set like NLSY97, with each cell representing hours worked for an individual in a particular year. The question is whether we can use math to split this data in a part that varies over time and a part that varies with the individuals, a more sophisticated version of what we did with the fixed effects model.

The simplest case is where there is no error term. This is called an **exact factor model**.

$$\mathbf{Y} = \mathbf{F}\mathbf{\Lambda} \tag{11.11}$$

where $\mathbf{F}$ is a $T \times K$ matrix and $\mathbf{\Lambda}$ is a $K \times N$ matrix. In this model, K represents the number of factors, $\mathbf{F}$ is the time-varying factor values and $\mathbf{\Lambda}$ is the set of factor weights that are unique to each individual.

One way to show that it is *not* possible to find a unique solution to Equation (11.11) is add a square $K \times K$ matrix, $\mathbf{A}$,

$$\mathbf{Y} = \mathbf{F}\mathbf{A}^{-1}\mathbf{A}\mathbf{\Lambda} \tag{11.12}$$

where the equation holds for any full-rank matrix $\mathbf{A}$. The implication is that many many factor values and factor weights are consistent with the observed data.

11.5.2 Convex Matrix Factorization

Are factor models useless?

Possibly. One solution is to assume that the weighting matrix is convex. That is, assume that all the values in the matrix are positive and rows sum to 1. Each individual in the data places weights on the factor values that are like probabilities. Note there is no compelling reason to make this restriction other than it is intuitive.

Assumption 8. $\sum_{k=1}^{K} \lambda_{ik} = 1$ *and* $\lambda_{ik} \in [0,1]$ *for all* $k \in \{1,...,K\}$ *and* $i \in \{1,...,N\}$.

Note that Assumption 8 is different from the convexity assumption proposed by Abadie et al. (2010). Here it is that each individual's weights on the factors sum to 1. In Abadie et al. (2010) it is that the treated individual places convex weights on the other individuals in the data. One does not imply the other.

The result of this restriction is that we can now uniquely find the matrix factors. Moreover, we may be able to determine exactly how many factors there are. The proof is a picture.

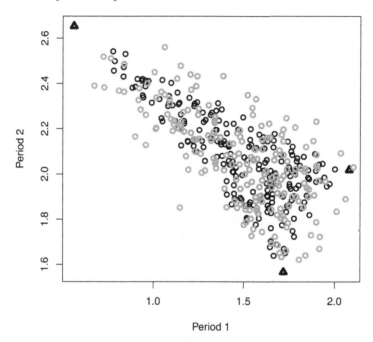

FIGURE 11.4
Plot of average outcomes by time period, with the factor vectors placed as triangles. The true outcomes are gray.

Figure 11.4 presents the outcomes (Ys) for the simulated data presented at the start of the chapter in two of the time periods. The black dots represent the results from an exact factor model. In the simulation this is $\bar{Y}$. Notice anything? We have plotted out a triangle. If we are able to observe the average outcomes then we have more than plotted out a triangle; we have identified the factor vectors. The factor vectors lie at the points of the triangle. We moved from an intractable problem first pointed out by Hotelling nearly 100 years ago, to a problem solved by plotting out the observed outcomes for two time periods.

Unfortunately, in our problem things are not as simple. Our data is represented by following **approximate matrix factorization** problem.

$$\mathbf{Y} = \mathbf{F}\mathbf{\Lambda} + \mathbf{E} \tag{11.13}$$

In the figure, the observed outcomes are represented by the gray circles. These make the triangle more rounded and make it difficult to factorize the matrix. Adams (2018) considers a restriction that there exists some independence of unobservables across time. Given this restriction, the paper shows that a mixture model can be used to identify the model.[5]

11.5.3 Synthetic Controls using Factors

If we can do the matrix factorization, we can create our synthetic controls. To do this, first note that we can write out the individual as a weighted average of the factor values.

$$\mathbf{y}_1 = \mathbf{F}\lambda_1 + \epsilon_1 \tag{11.14}$$

In addition, we can write the potential controls as a different weighted average.

$$\mathbf{Y}_{-1} = \mathbf{F}\mathbf{\Lambda}_{-1} + \mathbf{E}_{-1} \tag{11.15}$$

where $\mathbf{Y}_{-1}$ represents the outcomes for all the individuals that are not individual 1.

Note that both equations have the same matrix of factor values. This means that we can substitute the second into the first. However, first we need to solve for $\mathbf{F}$. To illustrate we will drop the unobserved term $\mathbf{E}$. Note because the weighting matrix $\mathbf{\Lambda}_{-1}$ is not square we use the same division idea presented in Chapter 1.[6]

$$\mathbf{F} = \mathbf{Y}_{-1}\mathbf{\Lambda}'_{-1}(\mathbf{\Lambda}_{-1}\mathbf{\Lambda}'_{-1})^{-1} \tag{11.16}$$

Substituting into Equation (11.14).

$$\hat{\mathbf{y}}_1 = \mathbf{Y}_{-1}\mathbf{\Lambda}'_{-1}(\mathbf{\Lambda}_{-1}\mathbf{\Lambda}'_{-1})^{-1}\lambda_1 \tag{11.17}$$

Thus, we can write the outcomes of individual 1 as a weighted average of the outcomes of the other individuals. This is our synthetic control.

11.5.4 Estimating the Weights

Equation (11.17) shows that in order to create our synthetic control we need to estimate the factor weight matrix ($\mathbf{\Lambda}$). Note that the matrix $\mathbf{\Lambda}_{-1}$ and the vector λ_1 are the constituent columns of the original matrix.

To make things doable we will assume that the unobserved characteristic in Equation (11.3) is normally distributed, $\epsilon_{it} \sim \mathcal{N}(0, \sigma^2)$. Given this assumption

---

[5]Mixture models are discussed in Chapter 12.

[6]Remember that with matrices, order matters.

we can use maximum-likelihood to factorize the matrix. We will do that in two steps.

Step 2. Find $\hat{\boldsymbol{\Lambda}}$. The second of the two steps is to determine the factor weights after we determine the factor value matrix ($\hat{\mathbf{F}}$) and the standard deviation of the unobserved characteristic ($\hat{\sigma}$).

$$\max_{\hat{\lambda}_i} \quad \sum_{t=1}^{T} \log\left(\phi\left(\frac{y_{it} - \hat{\mathbf{F}}_t'\hat{\lambda}_i}{\hat{\sigma}}\right)\right) \tag{11.18}$$

Similar to the Abadie et al. (2010) estimator, this is calculated for each individual i.

Step 1. Solve for $\hat{\mathbf{F}}$ and $\hat{\sigma}$. The first step also uses maximum-likelihood. However, things are more complicated because we don't know the weights. It is assumed these are distributed uniformly.

$$\max_{\{\hat{\mathbf{F}}_t, \hat{\sigma}\}} \sum_{t=1}^{T} \sum_{i=1}^{N} \log\left(\int_{l\in\mathcal{L}} \phi\left(\frac{y_{it} - \hat{\mathbf{F}}_t'l}{\hat{\sigma}}\right) dl\right) - NT\log(\hat{\sigma}) \tag{11.19}$$

where $\mathcal{L} = [0,1]^K$.

11.5.5 Convex Factor Model Estimator in R

Step 1. Estimate the factor value matrix, $\mathbf{F}$, and the distribution of the error term, σ.

```
> f_cmf <- function(par,Y,K,Reps=10) {
+    epsilon <- 1e-20
+    set.seed(123456789)
+    Y <- as.matrix(Y)
+    T <- dim(Y)[1]
+    N <- dim(Y)[2]
+    sigma <- exp(par[1])
+    F <- matrix(exp(par[2:(K*T+1)]),nrow=T)
+    p <- matrix(0,T,N)
+    for (r in 1:Reps) {
+       L <- matrix(runif(N*K),nrow=K)
+       L <- as.matrix(t(t(L)/rowSums(t(L))))
+       p <- p + dnorm((Y - F%*%L)/sigma)
+    }
+    p <- p/Reps
+    # small number added becuase of logging in the next step.
+    log_lik <- -(mean(log(p + epsilon)) - log(sigma))
+    return(log_lik)
+ }
```

We can test the estimator on the synthetic data described in the introduction. Note we will cheat and set the initial parameter values for the standard deviation equal to its true value. Again it is assumed we know the true number of factors. Note that this is estimated on just the first five time periods.

```
> par <- c(log(.1),rnorm(15,mean=0,sd=1))
> a <- optim(par=par,fn=f_cmf,Y=Ys[1:5,],K=3,
+               control = list(trace = 0,maxit=1000000))
> sigma_hat <- exp(a$par[1])
> sigma_hat

[1] 0.2138843

> F_hat <- matrix(exp(a$par[2:length(a$par)]), nrow=5)
```

The estimated value for the standard deviation of the error is not too far from its true value of 0.1. Table 11.1 presents the estimated factor values for the synthetic data. The true factor values lie between 0 and 3, so these estimates are a little off.

	1	2	3
1	3.60	0.17	0.01
2	2.13	1.92	2.13
3	1.54	0.97	1.71
4	3.55	0.13	0.68
5	2.84	1.91	1.85

TABLE 11.1
Estimated factor value matrix for the synthetic data created at the start of the chapter.

Step 2. Estimate the weighting matrix, Λ.

```
> f_lambda <- function(par, Y, F, sigma) {
+     epsilon <- 1e-20
+     Y <- as.matrix(Y)
+     F <- as.matrix(F)
+     L <- as.matrix(exp(par)/sum(exp(par)))
+     log_lik <- -sum(log(dnorm((Y - F%*%L)/sigma) + epsilon))
+     return(log_lik)
+ }

> Lambda_hat <- matrix(NA,3,200)
> for (i in 1:200) {
+     parti <- rnorm(3,mean=0,sd=20)
+     ai <- optim(par=parti, fn=f_lambda, Y=Ys[1:5,i],
+                 F=F_hat, sigma=sigma_hat)
+     Lambda_hat[,i] <- exp(ai$par)/sum(exp(ai$par))
```

```
+    #print(i)
+ }
```

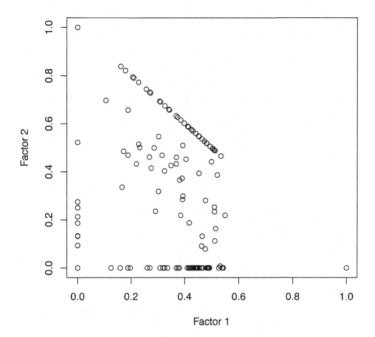

FIGURE 11.5
Plot of factor weights for the synthetic data.

Figure 11.5 presents a plot of the factor weights. The true values are more spread out than the estimates here.

Step 3. Once the factor model is estimated, the component parts can be used to determine the counter-factual outcome and the treatment effect.

```
> Ln1 <- Lambda_hat[,2:200]
> L1 <- Lambda_hat[,1]
> Yn1 <- Ys[31:35,2:200]
> y1_hat_fm <- Yn1%*%t(Ln1)%*%solve(Ln1%*%t(Ln1))%*%L1
```

Again, we can compare the predicted values to actual outcomes for individual 1. Figure 11.6 presents the actual and predicted outcomes with perfect prediction on the 45 degree line.

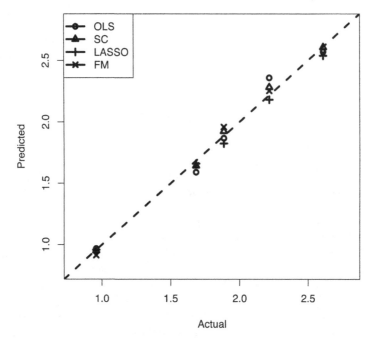

FIGURE 11.6
Plot of actual vs. predicted outcomes for individual 1 in periods 31 to 35. The 45 degree line is drawn and represents perfect prediction.

11.6 Returning to Minimum Wage Effects

In Chapter 10 we used a standard fixed effects model to analyze the impact of the 2007-2009 increase in the federal minimum wage on hours worked. The analysis follows the approach of Currie and Fallick (1996) but uses data from NLSY97.

11.6.1 NLSY97 Data

We can now return to the problem of estimating the effect of the federal minimum wage increase. In order to reduce the computational burden, the size of the treated and non-treated groups is substantially restricted.

```
> # using the same data as Chapter 10
> N2 <- 600  # restrict for computational reasons.
> T2 <- 18
> Y2 <- Y[,1:N2]
```

```
> W2 <- W[,1:N2]
> rate_07 <- W2[11,]/Y2[11,]
> # calculates the wage rate for each person.
> treat1 <- ifelse(rate_07<7.26 | W2[11,]==0 | Y2[11,]==0,1,0)
> treat <- matrix(0,T2,N2)
> for (i in 11:T2) {
+    treat[i,] <- treat1[1:N2]
+ }
> lm1 <- f_fe(Y2, treat)
> # fixed effects regression.

> Y2 <- matrix(NA,T2,N2)
> Y2[-lm1$na.action] <- lm1$residuals
> # na.action is an index of the NAs created in lm()
> # remember we can index a matrix as if it is a vector.
> Y_t <- Y2[,treat1==1]
> Y_t <- Y2[,is.na(colSums(Y_t))==0]
> Y_t <- Y_t[,1:5] # restrict for computational reasons.
> Y_nt <- Y2[,treat1==0]
> Y_nt <- Y_nt[,is.na(colSums(Y_nt))==0]
> # split into treated and non-treated groups
> # keep only those with data observed in all time periods.
```

11.6.2 Synthetic Control Estimates

```
> N_t <- dim(Y_t)[2]
> N_nt <- dim(Y_nt)[2]
> mat_treat <- matrix(NA,8,N_t)
> for (i in 1:N_t) {
+    pari <- rnorm(N_nt)
+    ai <- optim(par=pari, fn=f_sc, yt=Y_t[1:10,i],
+               ynt=Y_nt[1:10,],
+               control = list(trace=0,maxit=100000000))
+    omega <- exp(ai$par)/sum(exp(ai$par))
+    mat_treat[,i] <- Y_t[11:18,i] - Y_nt[11:18,]%*%omega
+    #print(i)
+ }
> mat_treat <- mat_treat + lm1$coefficients[2]
> # add back in the results of the fixed effects.
> plot_res_sc <- matrix(NA,8,3)
> for (i in 1:8) {
+    plot_res_sc[i,] <- quantile(mat_treat[i,],c(0.2,0.5,0.8),
+                               na.rm = TRUE)
+ }
> # quantiles of results at 0.2, 0.5, and 0.8.
```

The weights are estimated separately for each individual. Then the treatment effect is measured for each individual.

Figure 11.7 presents the distribution of the treatment effect for the five individuals analyzed. At least for these five people, the effect of the minimum wage increase is not large.

11.6.3 LASSO Estimates

The estimator goes through a number of different tuning parameters and then presents the results for the one that minimizes root mean squared error on the test data set.

```
> Y_tr <- Y_t[1:8,]  # training data
> Y_tt <- Y_t[9:10,]  # test data
> Y_ntr <- Y_nt[1:8,]
> Y_ntt <- Y_nt[9:10,]
> treat3 <- matrix(NA,8,dim(Y_t)[2])
> for (i in 1:dim(Y_t)[2]) {
+   omega_hat <- f_lasso(y_tr=Y_tr[,i],y_tt=Y_tt[,i],
+                 Y_ntr=Y_ntr,Y_ntt=Y_ntt,K=10)
+   treat3[,i] <- Y_t[11:18,i] -
+     cbind(1,Y_nt[11:18,])%*%omega_hat
+   #print(i)
+ }
> treat3 <- treat3 + lm1$coefficients[2]
> plot_res_ls <-matrix(NA,8,3)
> for (i in 1:8) {
+   plot_res_ls[i,] <- quantile(treat3[i,],c(0.2,0.5,0.8),
+                           na.rm = TRUE)
+ }
```

The results with this estimator are similar to those presented above. See Figure 11.7. You should check to see what happens when the set of tuning parameters is changed. The number of treated and non-treated individuals has also been restricted to reduce computational time.

11.6.4 Factor Model Estimates

The convex factor model approach is slightly different from that above.

Step 0. Set the data up. Note that the panel is aggregated up to half-decades. This is done for two reasons. First, it reduces the computational time for the problem. Second, it accounts for the possibility that there is **auto-correlation** in the annual data.[7] Ruling out auto-correlation is important for the identification argument presented in Adams (2018). It also allows more

---

[7] Auto-correlation refers to correlation in the unobserved characteristics over time.

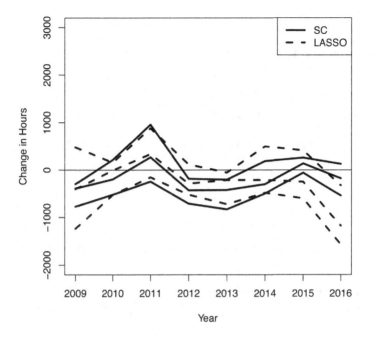

FIGURE 11.7
Plot of quantiles of the individual treatment effect for federal minimum wage change on five treated individuals in the NLSY97 data. The quantiles are the 20%, 50% and 80%.

factors to be observed.[8] However, the estimator uses most of the individual observations. We don't need to restrict it for computational reasons.

```
> Y_1t <- colMeans(Y[1:5,treat0==1], na.rm = TRUE)
> Y_2t <- colMeans(Y[6:10,treat0==1], na.rm = TRUE)
> Y_3t <- colMeans(Y[11:15,treat0==1], na.rm = TRUE)
> Y_1nt <- colMeans(Y[1:5,treat0==0], na.rm = TRUE)
> Y_2nt <- colMeans(Y[6:10,treat0==0], na.rm = TRUE)
> Y_3nt <- colMeans(Y[11:15,treat0==0], na.rm = TRUE)
> Y_t <- rbind(Y_1t,Y_2t,Y_3t)
> index_t_na <- is.na(colMeans(Y_t))==0
> Y_t <- Y_t[,index_t_na]
> Y_nt <- rbind(Y_1nt,Y_2nt,Y_3nt)
> Y_nt <- Y_nt[,is.na(colMeans(Y_nt))==0]
> Y_fm <- cbind(Y_t,Y_nt)
```

---

[8]Try re-doing without aggregation. The auto-correlation in the data will tend to imply that there are only two factors (Onatski and Wang, 2020).

```
> N_t <- dim(Y_t)[2]
> N_nt <- dim(Y_nt)[2]
```

Step 1. Estimate **F** and σ in the pre-treatment period. Note that we assume that there are three factors.

```
> par <- c(log(400),rnorm(6,mean=0,sd=10))
> a <- optim(par=par, fn=f_cmf,
+                Y=Y_fm[1:2,], K=3, Reps=100,
+                  control = list(trace = 0,maxit=1000000))
> # sigma
> sigma_hat <- exp(a$par[1])
> sigma_hat

[1] 411.6859

> F_hat <- matrix(exp(a$par[2:length(a$par)]), nrow=2)
```

Table 11.2 shows that there is a fair amount of variation in the hours.[9] The factors make points of a triangle at 0 hours for both time periods, 0 hours for period 1 and 4,000 hours for period 2, and 1700 hours for period 1 and 0 hours for period 2.

	1	2	3
1	0.10	1737.14	0.10
2	4024.42	0.01	0.00

TABLE 11.2
Estimated factor value matrix for the NLSY97 data. There are three factors and two time periods.

Step 2. Determine the weights.

```
> Lambda_hat <- matrix(NA,3,dim(Y_fm)[2])
> set.seed(123456789)
> for (i in 1:dim(Y_fm)[2]) {
+   ai <- optim(par=rnorm(3,mean=0,sd=1), fn=f_lambda,
+               Y=Y_fm[1:2,i], F=F_hat, sigma=sigma_hat)
+   Lambda_hat[,i] <- exp(ai$par)/sum(exp(ai$par))
+   #print(i)
+ }
```

Figure 11.8 shows that weights are spread out, perhaps unsurprising given that there are over 5,000 individuals.

---

[9]One issue with the estimator is that it will tend to run the estimate of the standard deviation down to zero if the starting value is too small.

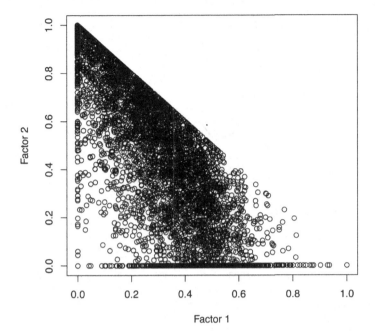

FIGURE 11.8

Plot of weights for the NLSY97 data. Note that there are three factors.

```
> Lhat_t <- Lambda_hat[,1:N_t]
> Lhat_nt <- Lambda_hat[,(N_t+1):dim(Lambda_hat)[2]]
> Omega_hat <- t(Lhat_nt)%*%solve(Lhat_nt%*%t(Lhat_nt))%*%Lhat_t
> itt <- Y_t[3,] - Y_nt[3,]%*%Omega_hat
> summary(as.vector(itt))

   Min. 1st Qu.  Median    Mean 3rd Qu.    Max.
-2314.8 -1263.9  -648.6  -529.5   124.8  3371.4
```

Figure 11.9 presents the individual treatment effect for the treated. It shows that a substantial proportion of individuals work fewer hours than they would have if they had not been treated by the increase in the federal minimum wage. That said, it is not clear that these people are worse off. In fact, the median treated individual had their income increase over $2,600 in the 5 years following the minimum wage increase.

```
> W_2t <- colMeans(W[6:10,treat0==1], na.rm = TRUE)
```

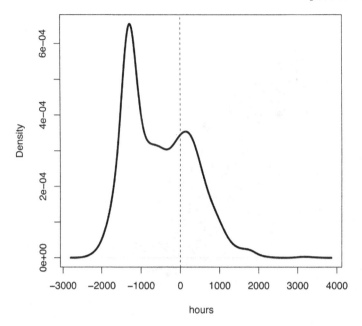

FIGURE 11.9
Density plot of the average loss in hours per year for the period 2008 to 2012.

```
> W_2t <- W_2t[index_t_na]
> W_3t <- colMeans(W[11:15,treat0==1], na.rm = TRUE)
> W_3t <- W_3t[index_t_na]
> inc <- W_3t - (Y_3t[index_t_na] - itt)*W_2t/Y_2t[index_t_na]
> median(as.vector(inc[is.na(inc)==0 & is.infinite(inc)==0]))

[1] 2599.427
```

11.7   Discussion and Further Reading

Panel data can be used to estimate the treatment effect by comparing outcomes before and after the treatment occurs. The problem is that time also affects the outcomes. In Chapter 10 this is solved by assuming that time has the same effect on the treated group as it does on the non-treated group. This chapter considers relaxing the assumption.

A synthetic control aims to construct a prediction of the counter-factual outcome. For the treated units, we observe their outcome after they have been treated, but not the outcome they would have had if they had not been treated. One way to construct the prediction is to use a weighted average of the outcomes for the non-treated units in the treated period. This is, in fact, what was done in Chapter 10. The difference here, is that pre-treatment period is used to find the appropriate weights for each treated unit.

The simplest and most obvious weights are the OLS weights. However, with only a few time periods and a large number of units, these weights are not identified. Abadie et al. (2010) suggests two restrictions, that the intercept is zero and the weights add to 1.

An alternative approach to restricting the model is to use a machine learning tool such as regularization. The chapter goes through a popular algorithm called LASSO. This is a least squares algorithm with an extra constraint that restricts the sum of the absolute value of the coefficients. The LASSO model tends to set coefficient values to zero. This leads to biased estimates, but potentially better predictions.

The chapter also shows that instead of making restrictions inferred by a particular factor model, we could simply use a factor model. Hotelling (1933) showed that the factor model is also not identified. The set of matrix factors that are consistent with the data is not unique. However, Fu et al. (2016) shows that restricting the factor weights to be positive and add to 1 (convex), it is possible to find unique matrix factors. The chapter presents a maximum-likelihood algorithm for finding the synthetic controls based on the convex factor model.

An excellent review of the various **synthetic control** techniques is Doudchenko and Imbens (2016). Guido Imbens, Susan Athey and co-authors have some review papers on using machine learning techniques in economics including in panel data models (Athey et al., 2019).

12

Mixture Models

12.1 Introduction

I am a colon cancer survivor. When I was diagnosed, I was told that had about an 75% probability of surviving 5 years (that was ten years ago!). My high survival probability is due to a combination of surgery and chemotherapy. Prior to the introduction of chemotherapy, the prognosis for someone with my diagnosis was 50%. The original chemos increased the 5-year survival to 70% and then newer chemos increased it to 75%. Am I alive because I had chemo?

I don't know.

The chemo may have done nothing. Prior to chemo, half the people with my diagnosis survived. I could have survived without chemo.

The example suggests that the treatment effect of chemotherapy is heterogeneous. For some, it has no effect, while for others it saves their life. There may even be a group of people that chemo harms. Imagine if we could determine which group each patient is in? Many fewer people would be exposed to the horrors of chemotherapy. Mixture models may be the answer.

Mixture models have become an important part of the econometrician's tool kit. However, in many cases they are not identified without strong and non-credible assumptions. Moreover, many researchers seem to be unaware that their results are relying on such non-credible assumptions.

This part of the book highlights how repeated measurement can be used to predict the effect of policies. This chapter considers more general cases than the panel data used in Chapters 10 and 11. The chapter uses the mixture model to solve the problem of measurement error. The chapter considers the problem of estimating returns to schooling using data from twins. It also returns to the question of whether an increase in the minimum wage will **cause** restaurants to decrease employment. The chapter highlights the great promise of mixture models, as well as the identification problems that arise.

12.2   Two-Type Mixture Models

Consider the simplest mixture problem. There are two **hidden types** which we will denote $\theta_t \in \{1, 2\}$. The observed outcome (y_t) depends on which is the true type.

$$y_t = \begin{cases} x_{1t} & \text{if } \theta_t = 1 \\ x_{2t} & \text{if } \theta_t = 2 \end{cases} \tag{12.1}$$

The density of y_t is the following mixture of two component densities.

$$f(y_t) = \Pr(\theta_t = 1)h_1(x_{1t}) + \Pr(\theta_t = 2)h_2(x_{2t}) \tag{12.2}$$

The question is whether we can determine the component parts of the mixture by observing the outcomes (y_t).

In my example above, the hidden types may represent how a particular colon cancer patient is affected by chemotherapy. Can we uncover the true effects (the x_t's) of the chemo by observing the aggregated outcomes, i.e. survival (y_t)?

12.2.1   Simulation of Two-Type Mixture

The simulation is based on the model presented above. Note that there are two observed outcomes $\{y_1, y_2\}$, both generated by an underlying set of hidden types. Also let $\Pr(\theta_t = 1) = \alpha$.

```
> set.seed(123456789)
> require(mvtnorm)
> N <- 10000
> mu <- c(0,10,-1,3)
> Sigma <- diag(1,4)*c(2,1,3,1)
> # diagonal matrix with different variances.
> x <- rmvnorm(N,mean=mu,sigma=Sigma)
> alpha <- 0.4
> ranN <- runif(N)
> y1 <- ifelse(ranN < alpha,x[,1],x[,2])
> y2 <- ifelse(ranN < alpha,x[,3],x[,4])
```

Looking at Figure 12.1 and the distribution of y_1, the question is whether we can back out the distribution of the x's from observing y_1. The answer appears to be yes! The distribution is bi-modal with the two modes at 0 and 10, which are the means of the distributions for each of the types. How about for the second case, the distribution of y_2? Can you make out the $h_1(x_1)$ and $h_2(x_2)$ densities? Maybe if you squint.

Simply observing the outcome may not be enough to discern the underlying distributions. If it was, the chapter would be very short.

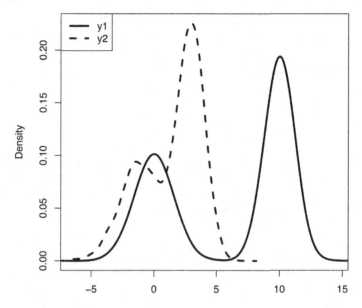

FIGURE 12.1
Density plots of two mixture distributions.

12.2.2 Knowing the Component Distributions

What if we knew the underlying type distributions? Could we back out the mixture weights? The answer is yes. To see this, consider the second mixture distribution broken into its component parts. In the simulated data we have the following relationship.

$$f(y_2) = \alpha h_3(y_2) + (1 - \alpha)h_4(y_2) \tag{12.3}$$

where $h_3(y_2) = \phi\left(\frac{y_2 - \mu_3}{\sigma_3}\right)$ and $h_4(y_2) = \phi\left(\frac{y_2 - \mu_4}{\sigma_4}\right)$. Noting that $\mu_3 = -1$, $\sigma_3^2 = 3$, $\mu_4 = 3$ and $\sigma_4^2 = 1$, we can rewrite as follows.

$$f(y_2) = \alpha\phi\left(\frac{y_2 + 1}{3^{0.5}}\right) + (1 - \alpha)\phi(y_2 - 3) \tag{12.4}$$

As we have a linear equation with one unknown (α), we can just rearrange to get the answer.

$$\alpha = \frac{f(y_2) - \phi(y_2 - 3)}{\phi\left(\frac{y_2 + 1}{3^{0.5}}\right) - \phi(y_2 - 3)} \tag{12.5}$$

At the median, we can calculate the density of the mixed distributions and use the formula presented in Equation (12.5). We get 0.43 where the true value is 0.4.

```
> q_y2 <- quantile(y2,0.5)
> a <- density(y2)$y[315]   # about the median
> b <- dnorm(q_y2,mean=3)
> c <- dnorm(q_y2,mean=-1,sd=3^(0.5))
> (a - b)/(c - b)
```

```
    50%
0.4248489
```

In some cases it is reasonable to assume the component distributions are known. In Appendix A, we analyze a statistical approach called **empirical Bayesian** estimation. This approach relies on the fact that we know the component distributions under random sampling.

12.2.3   Observing Multiple Signals

What if we don't know the component distributions? In general, we are back to square one.

We can make progress if we observe more than one outcome being generated by the same process. Consider the simulated data generated above. I didn't highlight it, but both y_1 and y_2 are two signals of the same underlying process. In one state, y_1 is determined by x_1 and y_2 is determine by x_3, in the other state y_1 is determined by x_2 and y_2 is determined by x_4.

We can back out the underlying mixtures by looking carefully at the distribution of one outcome conditional on the the second outcome. To see this, write out the probability of observing a certain value of y_1 given that $y_2 = a$.

$$
\begin{aligned}
f(y_1|y_2 = a) \ &= \Pr(\theta = 1|y_2 = a)f(y_1|\theta = 1, y_2 = a) \\
&+ \Pr(\theta = 2|y_2 = a)f(y_1|\theta = 2, y_2 = a) \\
&= \Pr(\theta = 1|y_2 = a)f(y_1|\theta = 1) + \Pr(\theta = 2|y_2 = a)f(y_1|\theta = 2) \\
&= \Pr(\theta = 1|y_2 = a)h_1(x_1) + \Pr(\theta = 2|y_2 = a)h_2(x_2)
\end{aligned}
$$

$$(12.6)$$

The simulated data satisfies an important assumption. Conditional on knowing the state, the value of y_1 is **independent** of the value of y_2. This assumption allows us to move from the first line to the second line. In math, we have $f(y_1|\theta = 1, y_2 = a) = f(y_1|\theta = 1)$. Once we know θ, the value of y_2 provides no additional information. This is a reasonable assumption if, as in the simulated data, the two observed outcomes are being generated by the same underlying process. They are **repeated measurements**.

The conditional probability is a mixture over the two component distributions. Importantly, the mixture weights can vary. The value of y_2 is dependent on the state; this means that the probability of the state changes conditional on the value of y_2. If there is a case where $\Pr(\theta = 1|y_2 = a) = 1$, then from Equation (12.6), we have $f(y_1|y_2 = a) = h_1(x_1)$. That is, we can infer $h_1(x_1)$. Similarly, if there is a case where $\Pr(\theta = 1|y_2 = a) = 0$ then we can infer $h_2(x_2)$.

Once we know these values, we can use the argument above to determine the mixing probabilities.

To illustrate, the next loop creates a matrix of conditional probabilities for our two simulated outcomes (y_1 and y_2). The loop breaks each outcome into a set of cells; then for each bucket of y_2 it calculates the probability of being in the various buckets for y_1.

Fu et al. (2016) point out that if these probabilities are precisely estimated then they will lie on a line between the two true underlying states. Moreover, the observed end points will correspond to the two true states. Identifying the model is simply a matter of plotting the data.

```
> min1 <- min(y1,y2)
> max1 <- max(y1,y2)
> K1 <- 10
> K2 <- 7
> epsilon <- 1e-5
> diff1 <- (max1 - min1)/K1
> diff2 <- (max1 - min1)/K2
> prob_mat <- matrix(NA,K2+1,K1+1)
> # conditional probability matrix
> prob_mat_x <- matrix(NA,K2+1,3)
> # actual hidden type conditional probability matrix
> low1 <- min1 - epsilon
> high1 <- low1 + epsilon + diff1
> for (k1 in 1:K1) {
+    low2 <- min1 - epsilon
+    high2 <- low2 + epsilon + diff2
+    prob_mat[1,1+k1] <- (low1 + high1)/2
+    y1_c <- y1[y2 > low1 & y2 < high1] # conditioning
+    for (k2 in 1:K2) {
+      prob_mat[1+k2,1] <- (low2 + high2)/2
+      prob_mat[1+k2,1+k1] <- mean(y1_c > low2 & y1_c < high2)
+      # probability conditional on 2nd signal
+      prob_mat_x[1+k2,2] <- mean(x[,1] > low2 & x[,1] < high2)
+      prob_mat_x[1+k2,3] <- mean(x[,2] > low2 & x[,2] < high2)
+      # probabilities conditional on actual hidden type.
+      low2 <- high2 - epsilon
+      high2 <- low2 + epsilon + diff2
+      #print(k2)
+    }
+    low1 <- high1 - epsilon
+    high1 <- low1 + epsilon + diff1
+    #print(k1)
+ }
> prob_mat_x[,1] <- prob_mat[,1]
```

Figure 12.2 plots the probabilities of two outcomes, one in the left tail and one in the right. These probabilities are denoted by the circles. Each probability is calculated given different values for y_2. The triangles are the probabilities derived from the underlying mixture distributions. In general, we don't observe the triangles, but the figure shows that the observed data enables us to determine where the triangles must lie.

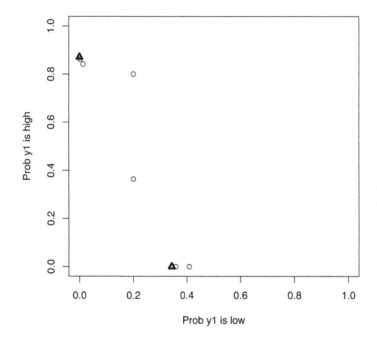

FIGURE 12.2
Plot of the probabilities $y_1 = -1.15$ on the x-axis and $y_1 = 12.06$ on the y-axis. The circles are the observed probabilities conditional on y_2 and the triangles are unobserved probabilities conditional on the true types.

Figure 12.2 shows that we can use the result of Fu et al. (2016) to identify the mixture model. If the conditional probabilities are estimated accurately then they must lie on the line between the two true types. If the conditional probabilities vary enough then the extremes will correspond to the case where only one type is true.

12.3 Two Signal Mixture Models

Below, we consider the measurement error problem associated with self-reported education levels. We are interested in comparing the difference in earnings for twins with different education levels. The problem is that most twins get the same level of education. This means that our analysis may be impacted by measurement error. It may be that all the differences we observe are due to reporting mistakes.

The proposed solution is to use two signals of the education level for each person in the study. Luckily, the twins were asked about both their own education level and the education level of their sibling. This means that for each person we have two signals of their true education level. One signal is the person's own report of their own education level. The second signal is the person's sibling's report of their education level.

12.3.1 Model of Two Signal Data

Consider a problem in which there are K hidden types which determine the values of two signals x_1 and x_2. In the twins study there are four types and the signals are the self-reported education level and the sibling-reported education level.

For simplicity assume that the type is represented by θ_k and the signals are a linear function of that value.

$$x_{i1} = a_1 + b_1\theta_k + \epsilon_{i1}$$

$$x_{i2} = a_2 + b_2\theta_k + \epsilon_{i2}$$

(12.7)

where $k \in \{1, ..., K\}$.

12.3.2 Simulation of Two Signal Data

In the simulated data there are four unknown types, say four education levels.

```
> set.seed(123456789)
> N <- 10000
> u <- c(1:4)
> p <- c(0.2,0.5,.25,0.05)
> ranN <- runif(N)
> theta <- ifelse(ranN < p[1], u[1],
+                 ifelse(ranN < p[2], u[2],
+                        ifelse(ranN < p[3], u[3], u[4])))
> e_1 <- rnorm(N,mean=0,sd=0.3)
> e_2 <- rnorm(N,mean=0,sd=0.3)
```

```
> x_1 <- 1 + .25*theta + e_1
> x_2 <- 0.5 + 0.1*theta + e_2
> summary(x_1)

   Min. 1st Qu.  Median    Mean 3rd Qu.     Max.
 0.2179  1.3847  1.7037  1.6980  2.0207   3.2098

> summary(x_2)

   Min. 1st Qu.  Median    Mean 3rd Qu.     Max.
-0.3676  0.5651  0.7897  0.7848  1.0046   1.9819
```

The two simulated signals have overlapping distributions. We can extend the ideas above to use these two signals to determine the unobserved types.

12.3.3 Conditional Independence of Signals

We can write down the joint density by the Law of Total Probability.

$$f(x_1, x_2) = \int_\theta h_1(x_1|x_2, \theta) h_2(x_2|\theta) g(\theta) \qquad (12.8)$$

We are interested in estimating the distribution of the hidden types $(g(\theta))$ and the conditional distributions $h_j(x_j|\theta)$. However, this time we only observe the joint distribution of the signals $(f(x_1, x_2))$. Is this enough? There is no magic here. We cannot determine h_1, h_2 and g from observing f.

However, if we are willing to assume that x_1 and x_2 are independent conditional on θ then we can rewrite Equation (12.8) in the following way.

$$f(x_1, x_2) = \int_\theta h_1(x_1|\theta) h_2(x_2|\theta) g(\theta) \qquad (12.9)$$

It is a subtle but important change. In the first case we cannot determine h_1, h_2 and g from observing f. Now we can, at least we can bound these equations following the arguments in Adams (2016) and Fu et al. (2016).

We have assumed that we have two signals and these signals are independent of each other conditional on the true underlying type. I think this assumption is reasonable for our twin's education reporting example. It implies that once we know the true education level for a person, then observing the sibling report provides no additional information about their own report.

12.3.4 Two Signal Estimator

The algorithm for solving the mixture model is suggested by Benaglia et al. (2009). The authors suggest simultaneously solving three equations.

$$\gamma(\theta|x_1, x_2) = \frac{h_1(x_1|\theta) h_2(x_2|\theta) g(\theta)}{\int_{\theta'} h_1(x_1|\theta') h_2(x_2|\theta') g(\theta')} \qquad (12.10)$$

The first is the posterior (γ) conditional on the observed outcomes.[1] This follows from Bayes' rule with conditional independence, so that $h(x_1, x_2|\theta) = h_1(x_1|\theta)h_2(x_2|\theta)$. The posterior is a function of the joint likelihood of observing the outcomes conditional on the true state, weighted by the prior.

$$g(\theta) = \int_{x_1, x_2} \gamma(\theta|x_1, x_2)f(x_1, x_2) \tag{12.11}$$

The second is the prior distribution which is calculated as the posteriors weighted by the probability of the two outcomes x_1 and x_2.

$$h(x_1, x_2|\theta) = \frac{\int_{x_2} \gamma(\theta|x_1, x_2)f(x_1, x_2)}{\int_{x_1, x_2} \gamma(\theta|x_1, x_2)f(x_1, x_2)} \tag{12.12}$$

The third is the likelihood function for the outcomes x_1 and x_2. This can be determined using Bayes's rule given the posterior and the observed distribution of the two signals.

Note that there may be multiple priors that satisfy these equations. There is no guarantee that there will be a unique solution. However, depending on the data, bounds around the solution set may be tight.

12.3.5 Two Signal Estimator Algorithm

Benaglia et al. (2009) suggest the following algorithm for solving the system.

Step 0. Let $t_{ik} \in \{0, 1\}$ and $\sum_{k=1}^{K} t_{ik} = 1$, denote the category that each observation i is assigned. Let $N_k = \sum_{i=1}^{N} t_{ik}$. The initial step assigns individual observations to one of four categories. In the estimator, this is done using kmeans().[2]

Step 1. Solve for the likelihood functions given the categorization.

$$\hat{h}_j(x|\hat{\theta}_k) = \frac{1}{N_k} \sum_{i=1}^{N_k} \mathbb{1}(x - \epsilon < x_{ij} < x + \epsilon)t_{ik} \tag{12.13}$$

Step 2. Solve for the prior.

$$\hat{g}(\hat{\theta}_k) = \frac{\sum_{i}^{N} t_{ik}}{N} \tag{12.14}$$

Step 3. Solve for the posterior.

$$\hat{\gamma}(\hat{\theta}_k|x_1, x_2) = \frac{\hat{h}_1(x_1|\hat{\theta}_k)\hat{h}_2(x_2|\hat{\theta}_k)\hat{g}(\hat{\theta}_k)}{\sum_{j=1}^{K} \hat{h}_1(x_1|\hat{\theta}_j)\hat{h}_2(x_2|\hat{\theta}_j} \tag{12.15}$$

---

[1] See Appendix A for a discussion of the use of Bayesian updating.
[2] kmeans() is a method for grouping correlated observations.

Step 4. Determine types. $t_{ik} = 1$ if and only if

$$\hat{\theta}_{ik} = \arg \max_{\theta_{ik}} \hat{\gamma}_i(\theta_{ik}|x_{1i}, x_{2i}) \qquad (12.16)$$

Go to Step 1.

Stop when the estimate of the prior (g) converges. Note that there is no guarantee that this algorithm will converge.[3]

12.3.6 Mixture Model Estimator in R

Mixture models can be estimated using npEM() from the package mixtools. It stands for non-parametric EM estimator, even though it is not actually an EM estimator.[4] The estimator below is based on the npEM() code. The original uses calls to C. This version uses **R**'s built in density() function.

```
> f_mixture <- function(X,centers=2,tol=1e-9,
+                               maxiter=500,verb=FALSE) {
+    X <- as.matrix(X)
+
+    # the function accepts either a matrix or just the number
+    # of hidden types.
+    if (length(centers) > 1) {
+      K <- dim(centers)[1]
+    } else {
+      K <- centers
+    }
+
+    # initial types
+    t <- kmeans(X, centers)$cluster
+    T <- matrix(NA,dim(X)[1],K)
+    for (k in 1:K) {
+      T[,k] <- ifelse(t==k,1,0)
+    }
+
+    # determine the initial prior estimate
+    g1 <- colMeans(T)
+
+    # loop to calculate the prior and posterior
+    diff <- sum(abs(1 - g1))
+    i <- 0
+    while (diff > tol & i < maxiter) {
+
```

---

[3]Levine et al. (2011) presents a similar algorithm which is guaranteed to converge.

[4]An expectation-maximization (EM) algorithm is a widely used alternative approach involving parametric assumptions.

```
+        # calculate the posterior
+        gamma <- t(matrix(rep(g1,dim(X)[1]), nrow=K))
+
+        # calculate the likelihood functions
+        for (k in 1:K) {
+          for (j in 1:dim(X)[2]) {
+            h <- density(X[t==k,j],from=min(X),to=max(X))
+            h$y <- h$y/sum(h$y)
+            gamma[,k] <-
+              gamma[,k]*sapply(c(1:dim(X)[1]), function(itr)
+                h$y[which.min((h$x - X[itr,j])^2)])
+          }
+        }
+        gamma <- gamma/rowSums(gamma)
+
+        # update the categorization of types
+        t <- max.col(gamma)
+
+        # update the prior estimate
+        g2 <- colMeans(gamma)
+
+        # check for convergence
+        diff <- sum(abs(g1 - g2))
+        g1 <- g2
+        i <- i + 1
+        if (verb) {
+          print(diff)
+          print(i)
+        }
+      }
+    return(list(g = g2, gamma = gamma, t = t))
+ }
> centers <- cbind(1 + 0.25*u, 0.5 + 0.1*u)
> # columns are number of signals (2 in this case)
> # rows are number of types (4 in this case.)
> a <- f_mixture(X=cbind(x_1,x_2),centers=centers)
```

We can use this model to determine the underlying types from simulated data. Figure 12.3 shows that the mixture model does a pretty good job of discerning the underlying types.

```
> mean(x_1[a$t==1])

[1] 1.190989

> mean(x_1[a$t==2])
```

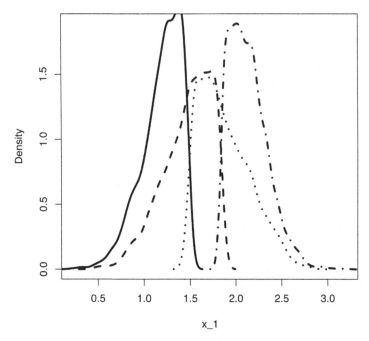

FIGURE 12.3
Density plots of the likelihood functions for each of the types conditional on x_1.

```
[1] 1.471282

> mean(x_1[a$t==3])

[1] 1.876143

> mean(x_1[a$t==4])

[1] 2.135286
```

The estimator finds that the hidden types are $\{1.19, 1.47, 1.88, 2.14\}$, which compares to the true values of $\{1.25, 1.5, 1.75, 2\}$.

12.4   Twins Reports and Returns to Schooling

The book returns to more or less where it began, with the question of measuring returns to schooling. The bane of the econometrician's efforts to estimate returns to schooling is "unmeasured ability." In two papers, Princeton trio Orley Ashenfelter, Alan Krueger and Cecilia Rouse suggest a unique solution - find a set of twins, ask them their education level and their wage and measure the differences in wages associated with the differences in education levels (Ashenfelter and Krueger, 1994; Ashenfelter and Rouse, 1998). This first-difference approach assumes that twins have the same unmeasured ability and differences it out.[5]

The problem is measurement error. Most twins have the same education level, so any observed difference in education level may not be real. Observed differences may be due to one or the other of the twins misreporting his education level. Such measurement error is likely to lead to estimates of returns to education that tend toward 0.

The section shows how a mixture model estimator can be used to estimate each person's true education level conditional on his own report and the report of his sibling. These true education levels are then used to estimate returns to schooling.

12.4.1 Mixture Model of Twin Reports

The mixture model approach solves the measurement error problem in two steps. In the first step, each individual's education level is assumed to be measured by two *different* signals. The first signal is the individual's own report of his education level. The second signal is the individual's sibling's report of his education level. Note that a sibling's report is likely to have different characteristics than an individual's own report.

To use the approach above we need to assume that the individual's own report is independent of their sibling's report of their education level, conditional on the true education level. This assumption seems quite reasonable. It implies that the reports only differ from the truth due to errors and mistakes.

Here, these two signals are used to estimate four "types" of individuals. Each type has a conditional distribution over education levels, where each is labeled for the mode of the conditional distribution of education level - 12, 14, 16 and 18.[6] These levels line up with the certificate granting years - high school diploma, associates degree, bachelor's degree and master's degree. To be clear, while the number of latent types is imposed on the estimated model, the conditional distributions (and their modes) are determined by the data.

The second step uses the results to estimate returns to schooling. It takes estimates from the mixture model to calculate the posterior probability of being a member of each education level type. These probabilities are used to weight the average education level for each type to give each person's estimated

---

[5]See discussion of first difference in Chapter 10.

[6]The fact that particular individuals may be poor at estimating both their own or their sibling's education level will not affect the procedure used here except through sampling correlation. See Ashenfelter and Krueger (1994) for discussion of this issue in regards to the IV approach.

education level - the "true" education level. Each twin is then matched with his sibling to calculate the difference in estimated true education levels. These differences are used in the place of the naive and IV estimated differences in the first difference regression models presented in Ashenfelter and Rouse (1998).

12.4.2   Twin Reports Data

The analysis uses the same data as Ashenfelter and Rouse (1998).[7]

```
> require(foreign) # package for reading older Stata data sets.
> x <- read.dta("pubtwins.dta")
> x_f <- x[is.na(x$first)==0,]  # first twin
> x_s <- x[is.na(x$first)==1,]  # second twin
> # educ "Twin1 Education";
> # educt "Twin1 Report of Twin2 Education";
> # educ_t "Twin2 Education";
> # educt_t "Twin2 Report of Twin1 Education";
> # the data is written out in pairs for each set of twins.
> lm1 <- lm(dlwage ~ deduc, data=x_f)
> lm2 <- lm(dlwage ~ deduc + duncov + dmaried + dtenure,
+             data=x_f)
```

Figure 12.4 shows two important takeaways from the data. First, for many twins there is no difference in the education levels. Second, there is a distinct positive relationship between differences in education level and income for the twins in the sample.

Ashenfelter and Krueger (1994) and Ashenfelter and Rouse (1998) present evidence that first differencing gives biased estimates. This bias is due to measurement error which is exacerbated by the first differencing. The authors' solution is to use an IV approach in which alternative reports are used as an instrument for the individual's own reports. The naive approach suggests an effect of 7% while the IV approach gives an estimate of 8.8%. Can you replicate these IV estimates?

12.4.3   Mixture Model Approach to Measurement Error

The IV approach suggested by Ashenfelter and Krueger (1994) makes very strong assumptions about the structure of the measurement error.[8] The mixture model approach suggests that the measurement error issue can be addressed without the need for the strong assumptions of IV. Note that both the IV

---

[7]The data is available here: https://dataspace.princeton.edu/jspui/handle/88435/dsp01rv042t084 or https://sites.google.com/view/microeconometricswithr/table-of-contents

[8]These assumptions are discussed in Chapter 3.

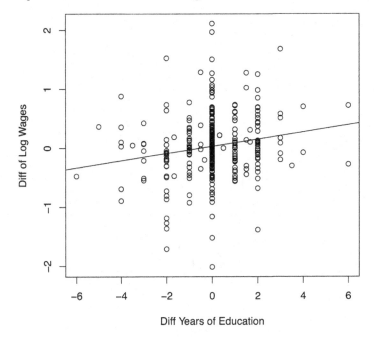

FIGURE 12.4
Wage differences by differences in education reports by twins. The line is the regression from above.

method and the mixture model method assume that reports are conditionally independent.

The first step is to determine the education level associated with each type. We need to split the data between "first" and "second" twins so that we can make the correct comparison.

```
> set.seed(123456789)
> y_f <- cbind(x_f$educ,x_f$educt_t)
> y_s <- cbind(x_s$educ,x_s$educt_t)
> y <- rbind(y_f,y_s)
```

The data is recombined to estimate the mixture model.

```
> centers <- matrix(rep(c(12, 14, 16, 18),2), nrow = 4)
> a <- f_mixture(y,centers,tol=1e-25)
```

Figure 12.5 presents the estimates of the likelihood of observing the twin's own report conditional on their true education level being 12, 14, 16 or 18. The plots show that for some people who report 13 years, their true education level is 12 years, while for others their true education level is 14 years.

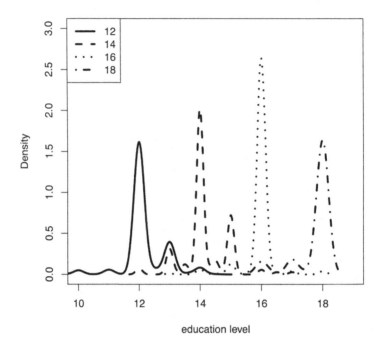

FIGURE 12.5
Density plot of the likelihood functions for observing the twin's own report conditional on their "true" education level.

The plots suggest that the model is *not* point identified. Adams (2016) shows that point identification depends on the existence of cases where the likelihood uniquely determines the type. The plot shows that the likelihoods have tails that are always overlapping. This means that the no likelihood is able to uniquely determine the type. That said, the existence of high likelihood ratios suggests that the identified set is tightly bound.

12.4.4   Estimating Returns to Schooling from Twins

Using the posterior estimate of each twin's "type" we can assign them and their sibling an education level. Then we can calculate the first-difference.

```
> ed <- c(12,14,16,18)
> x_f$deduc <- a$gamma[1:340,]%*%ed - a$gamma[341:680,]%*%ed
> # difference in "true" education levels
> # for each twin
> lm3 <- lm(dlwage ~ deduc, data = x_f)
```

```
> lm4 <- lm(dlwage ~ deduc + duncov + dmaried + dtenure,
+            data=x_f)
```

	Dependent variable:			
	dlwage			
	(1)	(2)	(3)	(4)
deduc	0.061***	0.073***	0.068***	0.082***
	(0.019)	(0.018)	(0.019)	(0.018)
duncov		0.086		0.085
		(0.056)		(0.055)
dmaried		0.024		0.049
		(0.073)		(0.073)
dtenure		0.023***		0.024***
		(0.003)		(0.003)
Constant	0.030	0.026	0.032	0.030
	(0.028)	(0.026)	(0.027)	(0.026)
Observations	340	333	340	333
R^2	0.031	0.170	0.036	0.178
Note:				*p<0.1; **p<0.05; ***p<0.01

TABLE 12.1
Results comparing the standard first difference estimates to the mixture model estimates. In each case the second model includes the standard explanatory variables from Ashenfelter and Rouse (1998). Models (1) and (2) are the first difference estimates, while models (3) and (4) are the respective mixture model estimates. The standard errors are reported in parentheses.

Table 12.1 presents the OLS results for log wages on the first-difference in education levels. The standard errors are reported in parenthesis. Note that these standard errors do not account for the mixture model estimation used in calculating the first difference in education level. These results can be compared to the first difference estimates in Ashenfelter and Rouse (1998). The mixture model estimates are somewhat higher suggesting that measurement error is in fact biasing down the first difference estimates. However, they are lower than the IV estimates (not presented).

12.5 Revisiting Minimum Wage Effects

I began the chapter by claiming that mixture models could be used to estimate heterogeneous treatment effects. This section tests that idea on Card
and Krueger's controversial minimum wage paper.[9] The original paper was
controversial for two reasons. First, it purported to show that a bedrock result
of economics wasn't true. Economic theory predicts that a minimum wage
increase *will* lead to a decrease in the demand for labor. The **difference in difference** results showed that the minimum wage increase in New Jersey caused
a (small) *increase* in the demand for labor. Second, the paper was used to
justify an increase in the federal minimum wage, a controversial policy.

A possible explanation for the result is that the change in the minimum
wage in New Jersey had heterogeneous effects. Economic theory is clear that an
increase in the cost of labor hours will lead to a decrease in demand for labor
hours. It is less clear on how that decrease will be achieved. Will restaurants
decrease the number of full-time staff or the number of part-time staff or
substitute from full-time to part-time or the other way around?

12.5.1 Restaurant Employment

The analysis uses the same data as Chapter 10, but with the mixture model
method described above. The analysis uses the pre-treatment period (period 1)
to estimate the hidden types. With these in hand, it uses the post-treatment
period (period 2) to estimate the difference in difference for each hidden type.
Note that the section assumes that the types do not change from period 1 to
period 2. The two signals of the restaurant's hidden type are the number of
full-time employees and the number of part-time employees.

```
> Y <- rbind(x3$FT[x3$TIME==1],x3$PT[x3$TIME==1],
+            x3$FT[x3$TIME==2],x3$PT[x3$TIME==2])
> Y1 <- Y[1:2,is.na(colSums(Y))==0] # period 1
> Y2 <- Y[3:4,is.na(colSums(Y))==0] # period 2
> treat  <- x3$STATE[x3$TIME==1]==1
> treat1 <- treat[is.na(colSums(Y))==0]
```

Table 12.2 presents summary statistics for the number of full-time and
part-time employees in the surveyed restaurants. Note that these firms tend to
have many fewer full-time employees than part-time employees.

---

[9]The ideas in this section are inspired by recent work by Jean-Marc Robin, an econometrician at Sciences Po in Paris.

	Full-Time	Part-Time
1	Min. : 0.000	Min. : 0.00
2	1st Qu.: 2.000	1st Qu.:11.00
3	Median : 6.000	Median :17.00
4	Mean : 8.202	Mean :18.84
5	3rd Qu.:11.250	3rd Qu.:25.00
6	Max. :60.000	Max. :60.00

TABLE 12.2
Summary table of the number of full-time and part-time employees in period 1.

12.5.2 Mixture Model Estimation of Restaurant Type

In the education example above there was a clear measure of the "hidden type," the years of education, particularly the diploma granting years. Here there is no clear indication. I proceed by *guessing* that there are firms that hire small, medium and large numbers of full-time employees and part-time employees. For full-time these are the groups centered at 2, 6 and 11. For part-time they are the groups centered at 11, 17 and 25. Moreover, there are some firms that hire a small number of both, some that hire a small number of one and a large number of the other. Lastly, through trial-and-error I paired down the initial groups to 5.[10]

```
> p1 <- c(2,6,2,2,6)
> p2 <- c(11,17,17,25,11)
> a <- f_mixture(X=t(Y1),centers=cbind(p1,p2),tol=1e-15)
> a$g

[1] 0.30725075 0.07859148 0.26042725 0.17348836 0.18024216
```

Figure 12.6 shows that there are more or less five types of restaurants. There are small restaurants that hire a small number of both types of employees. There are small restaurants that tend to hire part-time employees and ones that tend to hire full-time employees. Then there are large restaurants that either have more part-time employees or more full-time employees.

12.5.3 Heterogeneous Minimum Wage Effect

The model determines the hidden type for each restaurant. With this information, we can run the **difference in difference** estimator presented in Chapter 10 for each type.

---

[10]I re-ran the mixture model a number of times. I began with all nine types and each time I removed the type which were estimated to be the smallest.

```
> plot(Y1[1,][a$t==1],Y1[2,][a$t==1],col=1,pch=1,lwd=2,
+       xlim=c(0,60), ylim=c(0,60),type="p",
+       xlab="# Full-Time", ylab="# Part-Time")
> for (i in 2:5) {
+    lines(Y1[1,][a$t==i],Y1[2,][a$t==i],col=i,pch=i,
+          lwd=2,type="p")
+ }
> legend("topright",c("Type 1","Type 2", "Type 3", "Type 4",
+                      "Type 5"), col=c(1:5),pch=c(1:5))
```

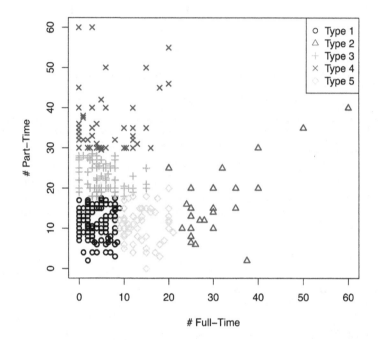

FIGURE 12.6
Plot of distribution of full-time and part-time employees by hidden type.

Figure 12.7 suggests that there was an effect of the increase in the minimum wage on small restaurants. There seems to be a shift down in the number of full-time and part-time employees that these firms hire after the minimum wage increases. The cloud disperses with some firms increasing employment, but relative to Pennsylvania firms there seem to be more firms with lower or zero levels of employment of part-time workers.

To see whether there is an effect for the small type we can estimate the difference and difference model for full-time workers. The estimate suggests

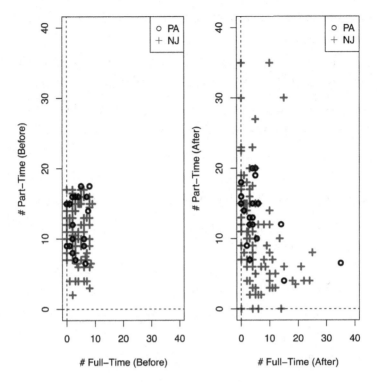

FIGURE 12.7
Plot of distribution of full-time and part-time employees for type 1 (small restaurants) before (left) and after (right) the minimum wage increase in New Jersey.

that the minimum wage increase reduced full-time employment by 17% for small restaurants.

```
> y_11 <- c(Y1[1,][!treat1 & a$t==1],Y1[1,][treat1 & a$t==1])
> # full-time employment for type 1 restaurants
> # prior to the minimum wage increase
> y_12 <- c(Y2[1,][!treat1 & a$t==1],Y2[1,][treat1 & a$t==1])
> # full-time employment for type 1 restaurants after
> # the minimum wage increase.
> treat = c(rep(0,length(Y1[1,][!treat1 & a$t==1])),
+           rep(1,length(Y1[1,][treat1 & a$t==1])))
> did1 <- f_did(rbind(y_11,y_12),treat)
> did1[3,3]/did1[1,1]

[1] -0.1710775
```

We can do the same analysis for part-time workers. Here the effect of the minimum wage increase is smaller, an 8% decrease in part-time workers for small restaurants.

```
> y_13 <- c(Y1[2,][!treat1 & a$t==1],Y1[2,][treat1 & a$t==1])
> y_14 <- c(Y2[2,][!treat1 & a$t==1],Y2[2,][treat1 & a$t==1])
> treat = c(rep(0,length(Y1[2,][!treat1 & a$t==1])),
+           rep(1,length(Y1[2,][treat1 & a$t==1])))
> did2 <- f_did(rbind(y_13,y_14),treat)
> did2[3,3]/did2[1,1]
```

[1] -0.08023572

The mixture model analysis finds that for type 1 restaurants (small restaurants) the minimum wage increase is associated with a large decrease in the number of full-time employees and a smaller but significant decrease in the number of part-time employees.

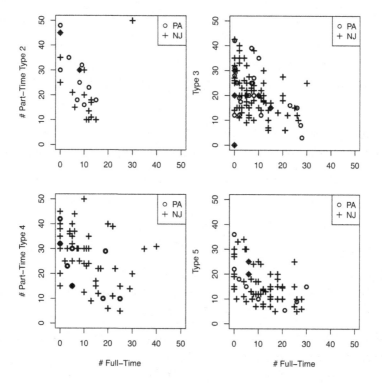

FIGURE 12.8
Plot of distribution of full-time and part-time employees for types 2 to 5 after the minimum wage increase in New Jersey.

Figure 12.8 presents the distribution of part-time and full-time employees by each of the other four hidden types. The figure shows that it is difficult to determine if there is any measurable effect of the minimum wage increase. In each case, the data is spread out in a cloud with no discernible decrease in employment. Can you replicate the difference and difference estimates for these types?

12.6   Discussion and Further Reading

Mixture models play an important role in microeconometrics. Unfortunately, mixture models are generally not identified without reliance on hard-to-justify parametric restrictions. The chapter considers the case where the econometrician observes multiple signals of the same underlying process. An example is reports of educational attainment. In this case we have two reports, one self-reported and the other, a report from the person's sibling. These two signals can be used to bound the distribution of the prior. In some cases the bounds can be tight.

Hu (2017) presents a comprehensive review of the three signal identification result and its applications. Haile and Kitamura (2019) limits the application to auctions with unobserved heterogeneity, but the paper provides an excellent overview of the various problems that occur in econometrics and a discussion of their solutions. The estimator presented here is based on Benaglia et al. (2009). While very nice to use, it is not guaranteed to converge. Levine et al. (2011) suggest a similar algorithm which does have nice convergence properties. This is a **majorization-minimization algorithm**.

The question of returns to schooling bookends this book. Chapters 1 and 2 presented OLS estimates using data from the NLSM 1966. The analysis suggested that one additional year of schooling increased wages by 7.8%. Card (1995) pointed out that this estimate is biased because there are unobserved characteristics that affect both the amount of schooling an individual receives and their income. The paper uses distance to college as an instrument and estimates that the returns to schooling is closer to 14%. Chapter 3 also presents LATE estimates that vary from 19% to 32%, although those estimates don't include other explanatory variables. These results suggest that there is a fair amount of heterogeneity in school returns. Chapter 6 returns to the question using the Heckman selection model and gets an average effect of 14%. The model estimates that there is variation in school returns and in fact returns are not positive for everyone. Chapter 9 uses GMM to re-estimate Card's IV approach with two instruments. It finds that returns are slightly smaller than suggested by OLS. This chapter uses reports from twins to estimate returns to schooling. After accounting for measurement error with a mixture model, the twins analysis suggests an average return of 8% for each additional year of schooling.

The chapter revisits Card and Krueger (1994) and their analysis of the minimum wage increase in New Jersey. The result suggests that minimum wage increase resulted in a substantial reduction in employment for small restaurants.

Part IV

Appendices

A

Measuring Uncertainty

A.1 Introduction

In microeconometrics it is important to provide information to the reader about how uncertain we are about the estimates. Standard microeconomic models of decision making suggest that people are risk averse and care about the uncertainty surrounding estimates (Kreps, 1990). Less standard, but increasingly common models suggest that decision makers care explicitly about the extent of the uncertainty (Klibanoff et al., 2005). How do we measure and report this uncertainty?

This chapter introduces the two main approaches used in statistics; classical statistics and Bayesian statistics. It may surprise you that statistics does not have a standard approach. Rather, statistics has two approaches that differ in important ways. The chapter discusses the problems and benefits of each. It also introduces a third way, called **empirical Bayesian statistics**. The third method marries classical and Bayesian statistics. The three approaches are illustrated by considering the question of whether a baseball player who played one game of major league baseball, John Paciorek, was better than the great Babe Ruth.

A.2 Classical Statistics

The approach you are probably most familiar with from economics is called **classical statistics**. Classical statistics uses the **analogy principle**. This principle states that if the econometrician is interested in some characteristic of the population, then she should use the analogy in the sample. Consider that we have a sample of data on player statistics for the WNBA. If we are interested in the average of height of WNBA players then we should use the average height of players in the sample. This is the sample analogy.

This section shows how the analogy principle can be used to provide information about the uncertainty surrounding our estimate.

A.2.1   A Model of a Sample

Consider a case where we observe a sample, $x = \{x_1, ..., x_N\}$, with N observations. In addition, assume that sample comes from a normal distribution, $x \sim \mathcal{N}(\mu, \sigma^2)$, where μ is the mean and σ^2 is the variance of the normal distribution.

In a typical problem we are interested in estimating the mean, μ. Let that estimate be denoted $\hat{\mu}$. We know that $\hat{\mu} \neq \mu$, but by how much do they differ?

A.2.2   Simulation of a Sample

Consider the simulation of a sample of size 10 drawn from a normal distribution with mean -3 and standard deviation of 5.

```
> set.seed(123456789)
> N <- 10
> mu <- -3
> sigma <- 5
> x <- rnorm(N, mean = mu, sd = sigma)
> dens_x <- density(x)
```

Figure A.1 shows that the sample distribution differs substantially from the true distribution. However, we are generally not interested in determining the true underlying distribution. Rather, we are interested in some aggregate characteristic of the distribution such as the mean.

```
> mean(x)
```

```
[1] -3.484872
```

```
> sd(x)
```

```
[1] 5.097328
```

The sample mean is -3.49, which is also substantially different from the true value of -3. The sample standard deviation is 5.10 which is not that different from the true value of 5.

How do we convey to the reader or the policy maker that our estimate of various statistics may not be accurate?

A.2.3   Many Imaginary Samples

One idea is to think about how our estimate would vary if we had a large number of different samples, all drawn from the same true distribution. For each imaginary sample that we draw from the true distribution, we will get a different estimate. As the number of samples gets large, we will have a distribution of the estimates. This distribution provides information about the uncertainty of our sample estimate. Luckily, with computers we can actually conduct this thought experiment.

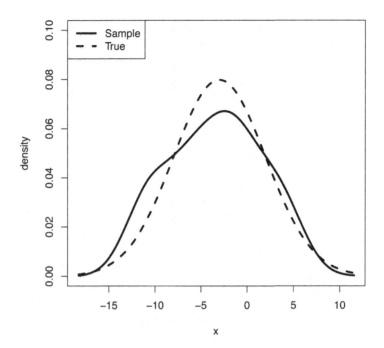

FIGURE A.1
Density plot of the sample and the true density function.

```
> M <- 500
> sample_est <- matrix(NA,M,N)
> for (m in 1:M) {
+    sample_est[m,] <- rnorm(N, mean = -3, sd = 5)
+ }
```

Consider the following simulated data. Figure A.2 presents the histogram of the distribution of sample means. The distribution is centered around the true mean of -3, but varies by more than 3 on each side. This illustrates that if we took a large number of samples our estimate would be correct on average.

We say that our estimate of the mean is **unbiased**. I generally don't find this particularly interesting. "Your estimate is almost certainly wrong but correct in expectation." Of course, it is also good to know if your estimate will have a tendency to under-estimate or over-estimate the true value.

More importantly, the weight of the distribution is around the true value. However, we cannot rule out the possibility of our mean estimate being as low as -6 or as high as 0. The extent of the dispersion is determined by the size of the sample. Try running the same experiment with a larger sample, say $N = 100$.

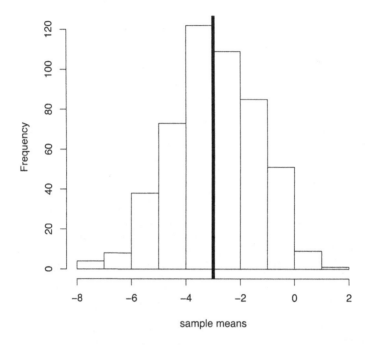

sample means

FIGURE A.2
Histogram of samples means where the true mean is -3.

If we were able to observe a large number of imaginary samples, we would be able to show how accurate our estimate is. But we don't observe any imaginary samples.

A.2.4 Law of Large Numbers

Another idea is to consider what happens if we have a large number of imaginary observations. If we had a very large sample our estimate of the mean would be very close to the true value. With 100,000 observations the sample mean is -3.001, which is quite close to -3. An estimator with this property is called **consistent**.

```
> set.seed(123456789)
> N <- 100000
> x <- rnorm(N, mean = -3, sd = 5)
> mean(x)

[1] -3.001334
```

Theorem 2. *Law of Large Numbers. If $\mathbb{E}(x_i) = \mu$ and $\mathbb{E}(x_i^2) < \infty$, then*

$$\lim_{N\to\infty} \frac{1}{N}\sum_{i=1}^{N} x_i = \mu \tag{A.1}$$

Various versions of Theorem 2 state that as the sample size gets large, the sample estimate converges (in some sense) to the true value. The Law of Large Numbers suggests that if our sample is "large enough" then our estimate may be close to the true value. Seems nice, but may not be that useful if our sample size is 10. It also does not provide us with any information about how uncertain our estimate actually is.

A.2.5 Central Limit Theorem

A central result to much of standard classical statistics is the Central Limit Theorem. The theorem states that as the number of observations gets large, then the estimated sample mean is distributed normally with a mean equal to the true mean and a standard deviation equal to the true standard deviation divided by the square root of the number of observations.

Theorem 3. *Central Limit Theorem. If $\mathbb{E}(x_i) = \mu$ and $\mathbb{E}(x_i^2) - \mu^2 = \sigma^2 < \infty$, then*

$$\lim_{N\to\infty} \sqrt{N}\left(\frac{1}{N}\sum_{i=1}^{N} x_i - \mu\right) \sim \mathcal{N}(0,\sigma^2) \tag{A.2}$$

The following is a simulation of the Central Limit Theorem. It presents the density of the distribution of estimated sample means as the size of the sample increases. Note that the distributions are normalized so that they will be standard normal (if the sample size is large enough).

```
> M <- 1000
> sample_means <- matrix(NA,M,4)
> for (m in 1:M) {
+    sample_means[m,1] <- mean(rnorm(10, mean = -3, sd = 5))
+    sample_means[m,2] <- mean(rnorm(100, mean = -3, sd = 5))
+    sample_means[m,3] <- mean(rnorm(1000, mean = -3, sd = 5))
+    sample_means[m,4] <- mean(rnorm(10000, mean = -3, sd = 5))
+ }
```

The interesting part of Figure A.3 is that even for small samples the density is close to a standard normal distribution.[1]

---

[1]We should be a little careful because we are using a normal distribution to approximate the density.

```
> plot(density((sample_means[,1]+3)*((10^(.5))/5)),
+       type="l",lwd=3,lty=1,xlab="x",main="")
> for (i in 2:4) {
+   lines(density((sample_means[,i]+3)*((10^(i*(.5))/5))),
+         lwd=3,lty=i)
+ }
```

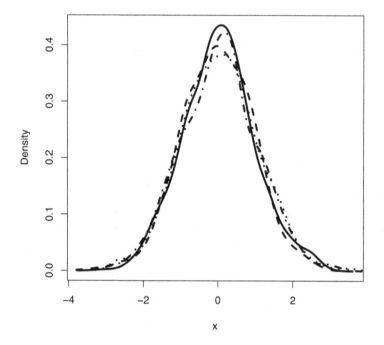

FIGURE A.3
Densities of samples means for various sample sizes with known true mean
and variance.

The Central Limit Theorem suggests a way of determining the uncertainty
associated with the estimate. It gives a distribution of the sample mean.
Moreover, the variance of the distribution, the uncertainty, is determined by
the true variance and the sample size. The uncertainty associated with the
estimate is smaller if the variance is smaller and if the sample size is larger.

The problem is that we don't know the variance. We don't know the
component parts of the distribution of the sample mean.

A.2.6   Approximation of the Limiting Distribution

The standard solution to not knowing the true values is to use an approximation
based on the analogy principle (Manski, 1990). Thus, we simply replace the

unknown true mean with its known analogy, the sample mean. We replace the unknown true variance with the known sample analogy, the sample variance.

Assumption 9. *Approximate Distribution.*

$$\hat{\mu} \sim \mathcal{N}\left(\hat{\mu}, \frac{\hat{\sigma}^2}{N}\right)$$

$$s.t.$$
$$\hat{\mu} = \frac{1}{N} \sum_{i=1}^{N} x_i$$

$$\hat{\sigma}^2 = \frac{1}{N} \sum_{i=1}^{N} x_i^2 - \hat{\mu}^2$$

(A.3)

Assumption 9 states that the estimated mean is distributed normally with a mean equal to itself and a variance equal to the sample variance divided by the sample size. This approximation gives us a measure of uncertainty that we can actually use.

There are two problems with this approach. First, there is no particularly good reason to believe that the sample mean is normally distributed. Sure, we know that it must be if the sample size is large, but our sample size is 10. Second, the measure of uncertainty is also estimated. Worse, it may be estimated poorly. Our measure of uncertainty is itself uncertain. The more uncertainty there is, the less certain we can be about how much uncertainty there is.

A.2.7 Simulation of Approximate Distributions

Consider five different samples and the corresponding approximate distribution of the sample mean from each.

```
> M <- 5
> N <- 10
> sample_means <- matrix(NA,M,2)
> x2 <- NULL
> for (m in 1:M) {
+    x1 <- rnorm(N,mean=mu,sd=sigma)
+    sample_means[m,] <- c(mean(x1),sd(x1)/(N)^(.5))
+    x2 <- c(x2,x1)
+ }
> x2 <- sort(x2)
```

Figure A.4 does not paint as rosy a picture as Figure A.3. The estimates of the distribution of the sample mean are all over the place. Our estimate of uncertainty is affected by how uncertain our estimates are.

```
> plot(x2,dnorm(x2,mean=sample_means[1,1],
+                sd=sample_means[1,2]), type="l",col="black",
+       lwd=3,ylab="density",xlab="x",
+       ylim=c(0,0.3), xlim=c(-10,3))
> for (m in 2:M) {
+    lines(x2,dnorm(x2,mean=sample_means[m,1],
+                sd=sample_means[m,2]), col="black",lwd=3)
+ }
```

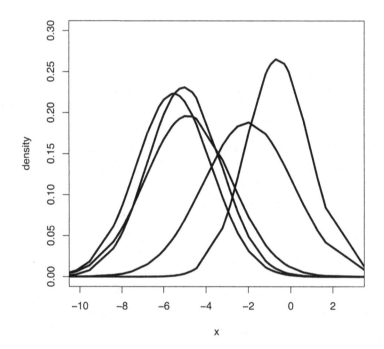

FIGURE A.4
Density plots of normal distributions using five different "analogies," where the
true mean is -3.

A.2.8   Bootstrap

An alternative approach to representing uncertainty, is called **bootstrapping**.
This idea was developed by Stanford statistician, Brad Efron. The idea takes
the analogy principle to its natural conclusion. If we need the true distribution
we should use its sample analog, that is the sample itself. Using a computer
we can create counterfactual pseudo-samples by re-drawing from the sample
with replacement.

Consider the following example using sample x_1.

```
> N <- 10
> x1 <- rnorm(N, mean=mu, sd=sigma)
> M <- 1000
> bs_mean <- matrix(NA,M,1)
> for (m in 1:M) {
+    index_bs <- round(runif(N,min=1,max=N))
+    bs_mean[m,1] <- mean(x1[index_bs])
+ }
> dens_bs <- density(bs_mean)
```

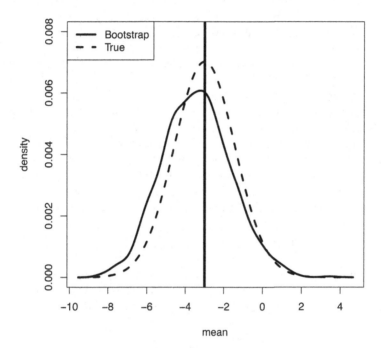

FIGURE A.5
Histogram of bootstrapped sample means where the true mean is -3 (vertical line).

Figure A.5 shows that our estimate of uncertainty is still wrong. The bootstrap measure of uncertainty no longer makes the arbitrary assumption that the sample mean is distributed normally. However, the small sample implies that we still have a poor estimate of the uncertainty.

A.2.9   Hypothesis Testing

A standard method for representing uncertainty is via an **hypothesis test**.
Above, the sample average is -3.49. Let's say it is important to know the
probability that the true mean is in fact 0 or is in fact less than 0. Can we
rule out the possibility that the true mean is greater than 0?

$$
\begin{aligned}
H_0 &: \mu \geq 0 \\
H_1 &: \mu < 0
\end{aligned}
\tag{A.4}
$$

Equation (A.4) presents a standard representation of the hypothesis that the
true mean is greater than or equal to 0. We call this the **null-hypothesis**. Note
that we are not testing whether the true mean is -3.49. Rather we are testing
whether it is *not* greater than 0. We are *ruling out* a possibility.

To test the hypothesis, we begin by assuming the hypothesis is true. That
is, we assume that the true mean is 0. Given that assumption, we can ask
whether it is reasonable that we would observe a sample average of -3.49. How
do we determine what is reasonable?

What is the probability that we could see a sample with a sample average
of -3.49 if the true mean is 0? To answer this question we need to know the
distribution of sample means. Above, we assumed that the distribution of
sample means is approximately normal with a mean equal to the true mean
and a variance equal to the true variance divided by the sample size. By
assumption, the true mean is 0. We don't know the true variance, but we can
use the analogy principle and replace it with the sample variance, which is
5.10.

$$
\Pr(\hat{\mu} = -3.49) = \phi \left(\sqrt{(10)} \frac{-3.49}{5.10} \right)
\tag{A.5}
$$

This will produce a small probability but that is partly because we are
asking about a point rather than a region. A better way is to ask at what
sample mean do we no longer think the hypothesis is true?

$$
\begin{aligned}
\Pr(\hat{\mu} &< t) = \alpha \\
\alpha &= \Phi \left(\sqrt{(10)} \frac{t}{5.10} \right) \\
&\text{and} \\
t &= \frac{5.10\Phi^{-1}(\alpha)}{\sqrt{(10)}}
\end{aligned}
\tag{A.6}
$$

Let α represent some small probability. It is the probability that $\hat{\mu}$ is less than
some **critical value** t.

```
> alpha = 0.05
> sd(x)*qnorm(alpha)/sqrt(10)

[1] -2.596642
```

The **critical region** is less than -2.60. In Figure A.6, this is represented

```
> pts = -5 + 10*c(1:1000)/1000
> plot(pts, dnorm(pts, sd=(5.10/sqrt(10))), lwd=3, xlab="x",
+       ylab="density", main="",type="l")
> abline(v=-2.597,lwd=3)
> abline(v=-3.49, lwd=3, lty=2)
```

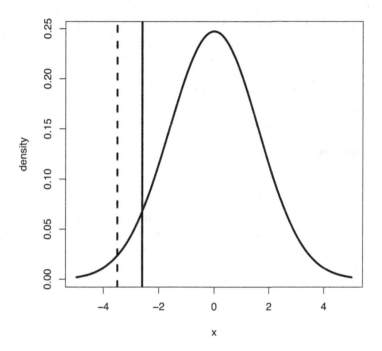

FIGURE A.6
Density of a normal distribution with standard deviation equal to $\frac{5.10}{\sqrt{(10)}}$. The critical value is -2.597 (solid line). Under the hypothesis of a mean of 0, the probability of observing a mean below the critical value is 0.05. The estimated mean is -3.49 (dashed line).

as the area under the curve to the left of the solid vertical line. Given the assumptions, we can reject the hypothesis that the true mean is 0 at the 5% level if the sample mean is less than -2.60, which it is.

An alternative is to calculate the **p-value**. This is the probability of observing a sample mean less than or equal to -3.49 given the hypothesis is true.

$$\Pr(\hat{\mu} < -3.49) = \Phi\left(\sqrt{(10)}\frac{-3.49}{5.10}\right) \tag{A.7}$$

```
> pnorm(sqrt(10)*((-3.49)/sd(x)))
```

[1] 0.01352641

This is 0.014 or 1.4%.

The nice property of the hypothesis test is that it allows the econometrician to assume the true mean, which is an unknown object. In addition, there are occasions where the null hypothesis is a value of interest. For example, Chapter 2 discusses the hypothesis that the direct effect of race on mortgage lending is 0. This particular hypothesis may have legal implications for bankers.

However, it is unclear why a hypothesis of 0 makes sense when describing the uncertainty around our estimate of -3.49. It may be preferable to present the 5th and 95th percentile of the approximate or bootstrap distribution. This is referred to as the **confidence interval**.

A.3 Bayesian Statistics

The major competitor to the classical approach is given the appellation "Bayesian" statistics (Berger, 1985). One should be careful about the names because "Bayesian" is also used to refer to the standard approach to decision making under uncertainty used in microeconomic theory. It may be best to think of them as two different approaches to decision making under uncertainty with the same name (Berger, 1985; Kreps, 1990).

A.3.1 Bayes' Rule

The Bayesian approach also provides us with a measure of the uncertainty but it uses Bayes' rule to do so. If we knew the set of possible true parameters, $\theta \in \Theta$, the likelihood of observing the sample, x, given the true parameters, $h(x|\theta)$, and the **prior** distribution over this set, $g(\theta)$, then we can use the following formula.[2]

$$\gamma(\theta|x) = \frac{h(x|\theta)g(\theta)}{\int_{\theta' \in \Theta} h(x|\theta')g(\theta')} \tag{A.8}$$

where $\gamma(\theta|x)$ is the **posterior** distribution. This tells us the probability distribution over the true parameters given the sample that is observed. By construction this posterior distribution provides information about the extent of the uncertainty of the estimate.

A.3.2 Determining the Posterior

To calculate the posterior, we need to know the prior and the likelihood functions. Assume we know the likelihood functions. Let's say we know the

---

[2]The prior distribution is the probability distribution over the true values prior to observing the data. The posterior distribution is distribution after observing the data.

possible set of distributions that generated the data. This may not be as crazy as it sounds. Consider the problem of determining a true probability. Say the probability that you will become King or Queen of England. Unless your name is Prince William, this is a pretty small number. Nevertheless it is a number between 0 and 1. I don't know the probability that you will become King or Queen of England but I do know that it is a number between 0 and 1. Here the problem is more complicated, but let's say that I know that the true mean must lie between -5 and 10. Also for simplicity, assume that I know that the true standard deviation is 5. So $\theta \in [-5, 10]$. And, I know that the true distribution is normal.

If I know the possible set of true means, the likelihood of observing the sample conditional on the true mean is just the likelihood function presented for OLS in Chapter 5. So we have the sample, the set of parameters and the likelihood function. All we need to use Equation (A.8) is the prior distribution.

What is the prior distribution? IDK. We could make an assumption that the prior distribution is uniform. In some cases, the uniform prior gives a sense of uncertainty around what the estimate could be.

A.3.3 Determining the Posterior in R

```
> log_norm <- function(x, mu, sigma) {
+    z = (x - mu)/sigma
+    return(sum(log(dnorm(z)) - log(sigma)))
+ }
> Theta <- 15*c(1:100)/100 - 10
> g <- rep(1/100,100) # uniform approximation
> h <- sapply(Theta, function(theta) log_norm(x1, theta, 5))
> gamma <- exp(h + log(g))/sum(exp(h + log(g)))
```

Figure A.7 presents the Bayesian posterior of the mean given the sample and a uniform prior. This posterior suggests that the estimate is quite uncertain with the weight of the distribution running between -7 and 0.

Given the assumptions of the Bayesian approach, the Bayesian approach is the "correct" and rational approach to presenting statistical results. It is even consistent with Bayesian decision theory.

However, the assumptions are where the rubber hits the road. The Bayesian approach makes very strong assumptions about what information the econometrician has available. Alternatively, it makes strong assumptions about things that the econometrician does not know. Moreover, these assumptions matter.

Neither the classical nor the Bayesian approach makes a lot of sense when data sets are relatively small. One uses non-credible information to make non-credible claims, while the other uses non-credible assumptions to make non-credible claims.

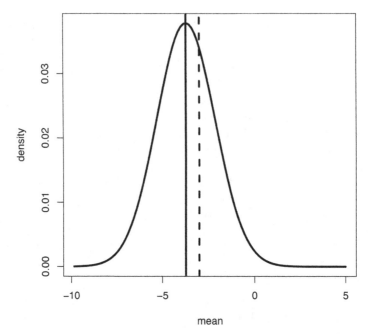

FIGURE A.7
Plot of Bayesian posterior distribution of mean estimate. The true value is -3 (dashed line). The analog estimate from the sample is lower (solid line).

A.4   Empirical Bayesian Estimation

In the middle of last century, Harvard statistician, Herb Robbins, suggested that there may exist a third way. He proposed an approach that used the logic of Bayesian statistics but replaced the strong assumptions with the classical idea of estimating the prior distribution using the analogy principle. He called this approach **empirical Bayesian**.

A.4.1   A Large Number of Samples

The empirical Bayesian approach may make sense when the econometrician has access to a large number of related samples. In panel data models, we can think of each time series for each individual as a separate but related sample. Similarly, we can think of each auction as a sample of bids from a large number of related samples.

Robbins pointed out that with a large number of related samples it may be possible to estimate the prior. That is, instead of assuming the prior, the data would provide an estimate of the prior. Given that the prior can be estimated, the rest of the procedure is Bayesian.

Robbins (1956) states that the observed distribution of sample means can be written as a mixture distribution, where a likelihood function is weighted by the true prior distribution.

$$f(x) = \int_{\theta \in \Theta} h(x|\theta)g(\theta) \tag{A.9}$$

where f is the observed distribution of sample means and $x \in \mathcal{X}$ is from the set of samples.

Robbins states that if the equation can be uniquely solved, then we have an estimate of the prior. In a series of papers, Brad Efron and coauthors present assumptions which allow the equation to be uniquely solved.

A.4.2 Solving for the Prior and Posterior

Given both f and h are known from Equation (A.9), Efron (2014) shows that under certain regularity conditions on h, we can uniquely solve for g. The proof uses linear algebra, but linear algebra doesn't actually help to find the solution.

Here we use an idea developed by statisticians at Penn State. It is similar to the mixture model estimator used in Chapter 11 (Benaglia et al., 2009). This estimator solves the following two equations iteratively. First, posit a prior, then use that to determine the posteriors.

$$\gamma_0(\theta|x) = \frac{h(x|\theta)g_0(\theta)}{\int_{\theta' \in \Theta} h(x|\theta')g_0(\theta')} \tag{A.10}$$

where γ_0 is the posterior given the posited prior g_0 for each $x \in \mathcal{X}$.

The posteriors can be aggregated up to provide a new estimate of the prior.

$$g_1(\theta) = \sum_{x \in \mathcal{X}} \gamma_0(\theta|x)f(x) \tag{A.11}$$

where g_1 is the next iteration of the estimate of the prior. With this prior we go back to Equation (A.10) and create a new estimate of the posterior.

We can say that the system converges if the following inequality holds.

$$\int_{\theta \in \Theta} |g_t(\theta) - g_{t-1}(\theta)| < \epsilon \tag{A.12}$$

for some small number ϵ, then $\hat{g}(\theta) = g_t(\theta)$ is the solution.

I am unaware of any proof showing that this method will actually lead to the solution, but if it does, then if conditions presented in Efron (2014) hold, it must be the unique solution.[3]

---

[3]Levine et al. (2011) shows that a similar algorithm does converge.

A.4.3   Solving for the Prior in R

Previously, we generated 500 imaginary samples in order to illustrate the idea that estimates of means have distributions. Now imagine we actually have a data set of these 500 samples. We don't know all the possible true means, but it seems reasonable to assume that the set of true means spans the observed data. We can approximate that set with 300 "true means" evenly spaced across the range. To make things simpler, assume that we know the mean is distributed normally with a standard deviation of 5. What we don't know is the true mean.

Note that we have 500 samples and we will calculate the posterior for each. We use a matrix with 500 rows and 300 columns.

```
> Theta = (max(sample_est) - min(sample_est))*c(1:300)/300 +
+    min(sample_est)
> H <- matrix(NA,500,300)
> for (i in 1:500) {
+    x <- sample_est[i,]
+    H[i,] <- sapply(c(1:300), function(k)
+       log_norm(x, mean(Theta[k]), 5))
+    #print(i)
+ }
```

Given the assumptions described above, we can calculate the likelihood of observing each sample given all the possible "true distributions." This provides an estimate of h from Equation (A.9).

The simplest place to start is to assume that the initial prior is a uniform distribution over the possible parameter values. This is similar to the assumption made for the Bayesian estimator. Here, it is only the starting point for estimating the prior.

```
> g0 <- rep(1/300,300)
> log_G0 <- log(t(matrix(rep(g0, 500),nrow=300)))
> gamma0 <- exp(H + log_G0)/sum(exp(H) + log_G0)
> g1 <- colSums(gamma0)/sum(gamma0) # new estimate of prior
```

Given this assumption and the likelihood calculated above, we have our first iteration of the estimate of the posterior. Integrating over the posterior estimates provides the next iteration of the prior.

We repeat the process to get a second iteration of the estimate of the prior. With these two we can then create a while () loop to determine when the process converges.

```
> log_G1 <- log(t(matrix(rep(g1, 500), nrow=300)))
> gamma1 <- exp(H + log_G1)/sum(exp(H) + log_G1))
> g2 <- colSums(gamma1)/sum(gamma1)
```

Figure A.8 presents the results of the first two iterations of the estimated

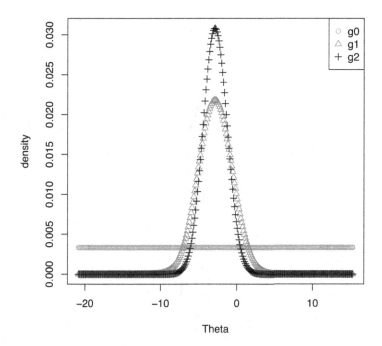

FIGURE A.8
Density of each of the first two iterations of the estimated.

priors. Note we start with a uniform or "flat" prior. The first iteration looks like a normal distribution. The second iteration is a normal distribution with thinner tails.

```
> epsilon <- 1e-3 # small number
> g_old <- g1
> g_new <- g2
> diff <- sum(abs(g_new - g_old))
> while (diff > epsilon) {
+    g_old <- g_new
+    log_G <- log(t(matrix(rep(g_old,500),nrow=300)))
+    gamma <- exp(H + log_G)/sum(exp(H + log_G))
+    g_new <- colSums(gamma)/sum(gamma)
+    diff <- sum(abs(g_new - g_old))
+    #print(diff)
+ }
```

The following is a method for creating a representation of the prior distribution. It creates a long vector where each sample mean is repeated in correspondence to its frequency draw from the prior.

```
> J <- 100000
> g_dist <- unlist(lapply(c(1:300), function(k)
+    rep(mean(Theta[k]),round(J*g_new[k]))))
```

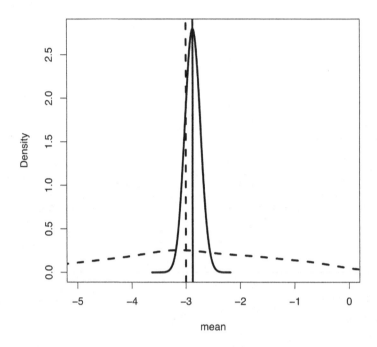

FIGURE A.9
Density of the estimated prior (solid line) and the observed sample means
(dashed line) and estimated prior mean is -2.88 (solid vertical line) where the
true mean is -3 (dashed line).

Figure A.9 presents the estimated density of the prior distribution for the
simulated data generated above. For comparison, it presents the density of the
observed sample means. Note that the empirical Bayesian prior is a lot tighter
than the observed distribution of sample means. This is a standard result
and is reason that the eBayesian estimator is called a **shrinkage estimator**. It
"shrinks" the estimates in toward the true mean.

A.4.4   Estimating the Posterior of the Mean

Once we have the prior estimated, we can use Bayes' rule to determine the
posterior distribution of the mean conditional on the observed sample. Consider
the sample of 10 observations generated above.

```
> h_x1 <-
+   sapply(c(1:300), function(k) log_norm(x1, Theta[k], 5))
> gamma_x1 <- exp(h_x1 + log(g_new))/sum(exp(h_x1 + log(g_new)))
```

Given this calculation of the posterior distribution, we can use the same procedure as above to create a density that can be plotted.

```
> J <- 100000
> g_dist_x1 <- unlist(lapply(c(1:300), function(k)
+   rep(Theta[k],round(J*gamma_x1[k]))))
```

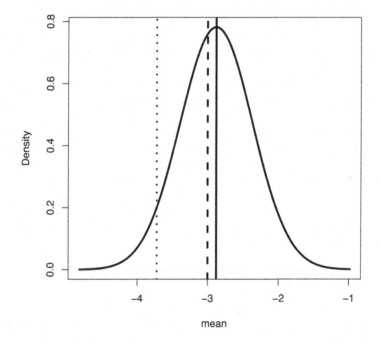

FIGURE A.10
Density of the estimated posterior for the sample where the true mean is -3 (vertical dashed line), mean of the sample is -3.72 (vertical dotted line) and mean of the eBayesian posterior is -2.88 (vertical solid line).

Figure A.10 presents the analog estimate which is close to -4, the true value which is -3 and the posterior estimate of the mean which is -2.88. Figure A.10 presents the posterior distribution using the eBayesian procedure. The fact that the posterior estimate is close to the true value is not really that interesting. It occurs because the sample has only 10 observations, so the posterior is mostly determined by the prior. The variance is a lot smaller than suggested by the classical procedures and even the standard Bayesian procedure.

A.5   The Sultan of the Small Sample Size

In baseball, the Sultan of Swing refers to Babe Ruth. Over his lifetime in major league baseball, the Bambino had a batting average of 0.341. That is, for every 1,000 "at bats," he hit the ball and got on base or scored a home run, 341 times.[4]

In 2013, the *New York Times* ran an article about John Paciorek (Hoffman, 2013). You have probably never heard of Paciorek, but if you have, you know that he has a lifetime batting average in major league baseball of 1.000. Paciorek had a "hit" with every "at bat," which totaled 3. Paciorek was a rookie who came up to the major leagues with the Houston Colt .45s for the last game of the 1963 season. He and his teammates had a great game against the hapless Mets. In the off-season, Paciorek was injured and never played another game in the majors.

We may ask whether Paciorek is "better" than Ruth because he has a higher lifetime batting average. You know the answer. Paciorek had only 3 at bats, while Ruth had thousands. If we think that each player has a "true" lifetime batting average, then the Law of Large Numbers suggests that Ruth's is probably pretty close to it. How close is Paciorek's? How uncertain is our estimate of Paciorek's lifetime batting average?

The section presents the standard classical and Bayesian approaches to representing the uncertainty around the estimate of John Paciorek's lifetime batting average. It shows that neither provides a satisfactory answer. It shows how to determine the empirical Bayesian posterior and shows that it does provide a satisfactory answer.

A.5.1   Classical or Bayesian?

Classical statistics gives a surprising answer. The analogy principle states that if we want to compare Paciorek to Ruth on their true lifetime batting average then we should use the lifetime batting average we observe. It says that Paciorek is better than Ruth, because $1.000 > 0.341$. Moreover there is little uncertainty. Using both the approximation method and the bootstrap method we get the same answer; there is no variance around the estimate of 1.000. Remember our sample data is $\{1, 1, 1\}$.

Bayesian statistics does slightly better. With only 3 at bats, the posterior distribution of Paciorek's lifetime batting average is completely dominated by the prior. What is the prior? Batting averages are numbers between 0 and 1, so one guess is a uniform distribution. However, anyone who knows anything about baseball will tell you that a uniform prior is ridiculous.

---

[4]An "at bat" refers to each time the batter goes up to bat, but does not include the times when the batter is "walked" and moves to first base without hitting the ball.

A.5.2 Uniform Prior

In addition to being a ridiculous assumption about baseball, the analysis in this section shows that it gives a ridiculous result.

Nevertheless, it is worth going through the exercise. As above, we calculate the likelihood function. Let all the true possible probabilities be the 999 numbers from $\frac{1}{1000}$ to $\frac{999}{1000}$. In this case the likelihood function is the binomial function. Note that this is not an assumption; it is really the binomial function.[5]

```
> f_bin <- function(N, p_hat, p) {
+    K = round(p_hat*N)
+    return(lchoose(N,K) + K*log(p) + (N-K)*log(1-p))
+ }
> # lchoose gives the log of the binomial coefficient.
> Theta <- c(1:999)/1000
> log_g <- log(rep(1/1000,999))
> h_jp <- sapply(Theta, function(theta) f_bin(3,1,theta))
> gamma_jp <- exp(h_jp + log_g)/sum(exp(h_jp + log_g))
> J <- 100000
> gamma_dist_jp <- unlist(lapply(c(1:999), function(j)
+      rep(j,round(J*gamma_jp[j]))))
> # mean
> sum(gamma_jp*Theta)

[1] 0.7995997
```

Figure A.11 suggests that Paciorek's lifetime batting average is very likely to be greater than Ruth's. In fact, the posterior probability is 0.986. However, this result is all driven by the assumption that the prior is uniform.

```
> sum(gamma_jp[342:999])

[1] 0.9863721
```

A.5.3 Estimating the Prior

To estimate the prior we need to observe a large number of lifetime batting averages. Luckily, thanks to people like Sean Lahman, we have access to data on almost every player to play Major League Baseball or Negro League Baseball between 1871 and 2018.[6] This is almost 20,000 players.

```
> M <- 1000
> x <- read.csv("bat_ave.csv", as.is = TRUE,nrows=M)
```

---

[5]Of course, there is an assumption. We are assuming random sampling. That is each at-bat is iid.

[6]Go to http://www.seanlahman.com/baseball-archive/statistics/ or https://sites.google.com/view/microeconometricswithr/table-of-contents

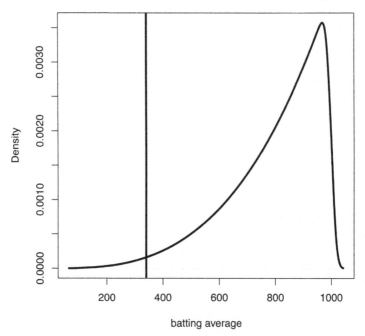

FIGURE A.11
Density of the estimated posterior for Paciorek's lifetime batting average given
a uniform prior. The vertical line is Ruth's lifetime batting average.

```
> # data set created from Lahman's data.
> # limited the observations for computational reasons.
> H <- matrix(NA,M,999)
> for (m in 1:M) {
+   H[m,] <- sapply(Theta, function(theta)
+     f_bin(round(x$AB[m]),x$AVE[m],theta))
+   #print(m)
+   # sapply uses apply on a vector.
+ }
> log_G <- t(matrix(rep(log_g,M),nrow = 999))
> gamma0 <- exp(H + log_G)/sum(exp(H + log_G))
> g1 <- colSums(gamma0)/sum(gamma0)
```

This uses the same method for estimating the prior as above.

```
> epsilon <- 0.001
> g_old <- exp(log_g)
> g_new <- g1
> diff <- sum(abs(g_new - g_old))
```

```
> while (diff > epsilon) {
+    g_old <- g_new
+    Log_G = log(t(matrix(rep(g_old,M),nrow=999)))
+    Gamma <- exp(H + Log_G)
+    Gamma <- Gamma/rowSums(Gamma)
+    g_new <- colSums(Gamma)/sum(Gamma)
+    diff <- sum(abs(g_new - g_old))
+    #print(diff)
+ }
```

A.5.4 Paciorek's Posterior

Using the likelihood for John Paciorek calculated in the previous section and the prior estimated above we can create vectors for graphing.

```
> gamma_jp2 <-
+    exp(h_jp + log(g_new))/sum(exp(h_jp + log(g_new)))
> J <- 100000
> g_dist_mlb <- unlist(lapply(c(1:999), function(j)
+    rep(Theta[j],round(J*g_new[j]))))
> # lapply uses apply on a list.
> # unlist turns a list into a vector.
> gamma_dist_jp2 <- unlist(lapply(c(1:999), function(j)
+    rep(Theta[j],round(J*gamma_jp2[j]))))
> # mean
> sum(gamma_jp2*Theta)
```

[1] 0.2506879

The mean of the posterior is 0.251. We could take this is our estimate of John Paciorek's true lifetime batting average. John's brothers Tom and Jim finished their MLB careers with lifetime batting averages of 0.282 and 0.228 respectively.

Figure A.12 presents the density functions of the estimated prior and the new posterior distribution for John Paciorek. The prior using data from almost 150 years of baseball suggests that the probability of any player having a lifetime batting average of 0.341 or greater is remote. The probability that John Paciorek's innate lifetime batting average is greater than 0.341 is 0.005, which is a lot smaller than both 1.000 and 0.986.

```
> sum(gamma_jp2[342:999])
```

[1] 0.004715468

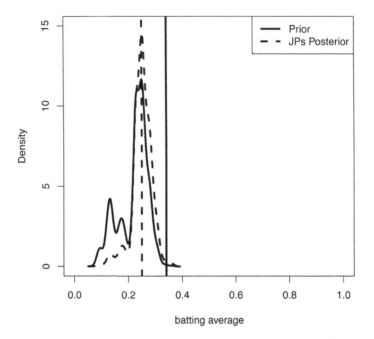

FIGURE A.12
Density of the estimated posterior (dashed line) for Paciorek's lifetime batting
average given an empirical Bayesian prior (solid line). The mean of the posterior
is the dashed vertical line. The sold vertical line is Ruth's lifetime batting
average.

A.6 Decision Theory

Statistics is fundamentally about making decisions under uncertainty. In some
cases the statistician herself has a decision problem. Should I collect more data
given data collection is costly and the data I have provides some information?
In other cases, the statistician acts as a go-between. They sit between the
data and the ultimate decision maker. The statistician provides a summary of
data to the decision maker. But how should she summarize the data? What
information should the decision maker get?

   Consider the problem above. If you are limited to providing one number
about John Paciorek's batting average, which number would you choose?
Consider a more realistic example. You work for a baseball team and you are
considering a trade of one first baseman for another. Assume, unrealistically,
that the team-manager only cares about batting average. If you are to provide

a single number to the team-manager that summarizes the information about the two players, which number does that?

A.6.1 Decision Making Under Uncertainty

Economists have been thinking about how to make decisions under uncertainty for a very long time. The standard paradigm is called expected utility theory. In this model there is some outcome x, say hits from our baseball example, and a probability distribution p, which is the probability of a hit. The probability of some outcome x_k is then p_k, $\Pr(x = x_k) = p_k$.

$$U(x, p) = \sum_{k=1}^{K} p_k u(x_k) \tag{A.13}$$

If there are K possible outcomes then the decision maker's utility is represented as expectation over the utility the decision maker gets from the hit. Note that if $u(x_k) = x_k$ then expected utility is equal to the average. In our baseball case, it is the batting average. More generally, it is not equal to the average.

The problem with this standard economic approach for our purposes is that we don't know what p is. In statistics, we have a sample estimate of p, $\hat{p}$ or a posterior distribution $\gamma(p)$ or an estimate of that distribution $\hat{\gamma}(p)$.

A.6.2 Compound Lotteries

One response to the fact that we don't know p is to introduce another distribution q, which represents the probability of p. The probability of some distribution p_l is q_l, $\Pr(p = p_l) = q_l$. In the baseball example above, p_l represents batting average (the probability of a hit) and q_l is probability that a particular player has that batting average.

If we don't know p but we do know q then we can write down expected utility as a **compound** expected utility.

$$U(x, p, q) = \sum_{l=1}^{L} \sum_{k=1}^{K} q_l p_{lk} u(x_k) \tag{A.14}$$

where there are L possible distributions of p and K possible outcomes. This representation uses something called **reduction of compound lotteries**. This means that the two vectors of probabilities are multiplied together to give one reduced vector of probabilities.

For most economists, this is generally where they stop. In fact, some argue that a rational decision maker must use Equation (A.14) as a basis for decision making under uncertainty. The problem with the representation in Equation (A.14) is that it does not really account for uncertainty over what p is.

It is easiest to see the implication if we assume that $u(x_k) = x_k$. In this case, Equation (A.14) implies the decision maker only cares about the average of the average. For a classical statistician, the average of the average is generally the analog estimate of the average. It is the number 1.000, for the case of John Paciorek. For a Bayesian statistician, it is the mean of the posterior distribution which is 0.800 for the standard Bayesian with a uniform prior and 0.251 for the eBayesian. The eBayesian is the only one of the three who would "rationally" prefer Babe Ruth to John Paciorek.

This is pretty strange given we have the whole field of statistics dedicated to determining the best way to measure uncertainty over p. Even economists spend a good chunk of time writing and talking about the uncertainty over p. Why spend so much time talking about something a rational decision maker cares naught for?

A.6.3   What in the World Does Wald Think?

In the 1940s, Hungarian-American mathematician, Abraham Wald, took decision theory directly to the statistician's problem. In Wald's set up, the decision maker has sample data and must make a choice given their utility. Interestingly, Wald considered the problem as a **game** with two players, Nature and Statistician. Nature chooses the true state of the world, although Nature could have **mixed strategies** (a distribution over the true state of the world). The Statistician does not know Nature's choice, although she may know the mixed strategies (Wald, 1949).[7]

Wald states that the Statistician should optimize against Nature by considering expectations over a utility function of the true state of the world and the outcome the Statistician receives.

$$U(x, p, q) = \sum_{l=1}^{L} q_l L(p_l, x) \tag{A.15}$$

where L is for "loss" gives the decision maker's utility given the true state of the world p_l and the set of outcomes x. In Classical statistics, q_l is determined by the approximation to the limit distribution (Assumption 9). In Bayesian statistics, it is the posterior distribution.

Importantly, if the "loss" function L is not linear then we no longer have **reduction of compound lotteries**. Our utility can now explicitly account for the uncertainty.

---

[7] A more modern game-theoretic approach would use the Nash equilibrium and set it up as a game of incomplete information. That is, the Statistician does not know Nature's type. Each type plays an equilibrium strategy in choosing the true state of the world. The Statistician does not know Nature's type, however by interacting with Nature through experiments, the Statistician can learn Nature's type. In Bayes-Nash equilibrium, the Statistician uses Bayes' rule to update the distribution over Nature's type and the true state of the world.

A.6.4 Two Types of Uncertainty?

In the discussion above there are two probability distributions, p and q. Some people use different words for the two types of uncertainty. The first is often called "risk" and the second is called "ambiguity." The first is the probability distribution over outcomes; it is the player's true batting average. The second is a distribution of Nature's choice of the state of the world. In Bayesian statistics, it is the posterior distribution. Can we have two types of uncertainty? Should they be treated differently?

Yes. The first type is fixed, albeit unknown. It is Nature's choice. The second type is not fixed. The second type of uncertainty varies with data and information. It may even be under control of the Statistician. For example, the Statistician may have the ability to create more data.

If we are to give just one number to summarize the data it should explicitly account for uncertainty associated with the sample data. For example the utility representation of Klibanoff et al. (2005),

$$U(x, p, q) = \sum_{l=1}^{L} q_l h \left(\sum_{k=1}^{K} p_k u(x_k) \right) \tag{A.16}$$

where h may be some non-linear function. With this function we can help the decision maker by providing summaries of the data that explicitly account for the uncertainty of our estimates.

A.7 Discussion and Further Reading

Given that much of the focus of statistics is providing information about uncertainty, there is surprisingly little agreement about the best approach. Moreover, the two leading candidates, classical and Bayesian statistics, do a surprisingly poor job at representing the uncertainty.

I find the empirical Bayesian approach suggested in Robbins (1956) and developed by Brad Efron and coauthors to be a compelling third way. Admittedly, the applicability of the approach may be limited, but I come across more and more applications every day. The approach also seems to be consistent with modern ideas on rational decision making under uncertainty. See Klibanoff et al. (2005) and Gilboa (2013).

Here, the eBayesian approach is illustrated with strong assumptions on the distribution functions. Efron (2014) points out that if the set of outcomes is finite then under random sampling the likelihood function is known; it is the **multinomial** function. Note, in the baseball example the likelihood function is a binomial function under random sampling.

B

Statistical Programming in **R**

B.1 Introduction

R is based on the language S, which was developed for statistical analysis by the same people who brought you C. When coding in any language it is important to understand how the language works. How it thinks.

R thinks that everything is a vector. Vectors are one dimensional "list-like" objects. They are the heart and soul of **R**. **R** can analyze vectors with amazing speed. A vector is not a list and not a matrix and it can only contain one type of value (numbers or strings, etc.) Lists can generally contain multiple types of objects, including other lists. Matrices are generally two dimensional objects that contain numbers.

In contrast to **R**, **Stata** thinks everything is a flat data set. It is optimized for operations involving columns of data. **Matlab** is a matrix based language, while **Python** and **Mathematica** are based on **LisP** (List Processing).

The chapter discusses the characteristics of objects including vectors, matrices and lists. It discusses basic control functions, statistics and standard optimization.

B.2 Objects in R

This section describes how **R** uses basic objects like vectors, matrices, lists and data frames. It also discusses manipulation of objects including mathematical operations.

B.2.1 Vectors

Consider the following operations. Create a vector of numbers called a, and a vector of words called b. You can try adding them together. The best way to do that is to use the c() function. This function concatenates or joins vectors together. Note what happens to the numbers when you do this.

```
> a <- c(1,2,3)
> # "<-" means assign which has a different meaning than "="
> b <- c("one","two","three")
> a

[1] 1 2 3

> b

[1] "one"    "two"    "three"

> # a + b
> # you cannot use a numerical operator like "+" on a
> # non-numeric object.
> d <- c(a,b)
> d

[1] "1"      "2"      "3"      "one"    "two"    "three"

> # note that the d is a vector of strings.
```

There are a couple of **R**-centric coding features. It is important to use <- when assigning a variable name. It is possible to use = but it doesn't always mean the same thing. Also note that c(1,2) is a vector with two elements, both numbers, while c("1","2") is a vector with two elements, both characters.

R can manipulate vectors very quickly. When coding it is important to remember this.

```
> b <- c(4, 5, 6)
> a + b

[1] 5 7 9

> a*b

[1]  4 10 18

> a/b

[1] 0.25 0.40 0.50

> a + 2

[1] 3 4 5

> a*2

[1] 2 4 6
```

```
> a/2
```

```
[1] 0.5 1.0 1.5
```

Check out the operations above. Look carefully at what happens in each case. In particular, note that each operation is cell by cell. Also note what happens when you have a mathematical operation involving a single number and a vector. The single number operates on each element of the vector.

```
> b <- c(4,5)
> a + b
```

```
[1] 5 7 7
```

```
> a*b
```

```
[1]   4 10 12
```

Note that things get a little strange when the two objects are not the same length. Importantly, **R** may do the operation regardless! It may or may not give you a warning!! Can you work out what it actually did?[1]

B.2.2 Matrices

R allows researchers to use a fairly large array of objects, which is very nice but it leads to some issues if you are not careful. Matrices are one such object and they can be very useful. In creating matrices we can use "cbind" which joins columns together or "rbind" which joins rows together. Note that the objects being joined must be the same length.

```
> a <- c(1,2,3)
> b <- c(4,5,6)
> A <- cbind(a,b)
> B <- rbind(a,b)
> A
```

```
      a b
[1,] 1 4
[2,] 2 5
[3,] 3 6
```

```
> B
```

```
    [,1] [,2] [,3]
a    1    2    3
b    4    5    6
```

---

[1]In the case of a + b, it did the following operations, 1 + 4, 2 + 5, 3 + 4. That is, **R** simply started over again with the first element of b. This is referred to as **recycling**.

```
> is.matrix(A)

[1] TRUE

> t(A)

  [,1] [,2] [,3]
a   1    2    3
b   4    5    6
```

The transpose operation is t() in **R**.

As with vectors, arithmetic operations on matrices are cell by cell. However, the matrices must be the same dimension in order to do cell by cell operations. Try the operations that are commented out below (remove the #).

```
> C <- A + 2
> A + C

     a  b
[1,] 4 10
[2,] 6 12
[3,] 8 14

> D <- B*2
> B*D

   [,1] [,2] [,3]
a    2    8   18
b   32   50   72

> # A*B
> # A + B
> A^2

     a  b
[1,] 1 16
[2,] 4 25
[3,] 9 36

> A + t(B)

     a  b
[1,] 2  8
[2,] 4 10
[3,] 6 12
```

You can also do standard matrix multiplication using the operator %*% to distinguish it from cell by cell multiplication. This operation follows the mathematical rules of matrix multiplication. For this operation to work, the two "inside" dimensions must be the same. Below, we are multiplying a 3×2 matrix **A** by a 2×3 matrix **B**. The operation **AB** works because the inside dimension is 2 for both. How about the operation **BA**? Try it below.

```
> A%*%B

     [,1] [,2] [,3]
[1,]   17   22   27
[2,]   22   29   36
[3,]   27   36   45

> # B%*%A
> # t(B)%*%A
```

B.2.3 Lists

The first computer language I really learned was called Logo. The language was developed by Seymour Papert in MIT's Artificial Intelligence Lab. Like its antecedent, Scratch, Logo was designed to help children learn mathematics and programming. Logo is based on LisP. My father, who was a computer scientist, would get very excited about the list processing ability of Logo. As a ten year old, I didn't quite understand the fuss. Today, using the list processing features of **R**, I am wistful of Logo's abilities. **R** is not a list based language but it can process lists.

```
> a_list <- list(a,b)
> a_list

[[1]]
[1] 1 2 3

[[2]]
[1] 4 5 6

> b_list <- list(a_list,A)
> b_list

[[1]]
[[1]][[1]]
[1] 1 2 3

[[1]][[2]]
[1] 4 5 6
```

```
[[2]]
     a b
[1,] 1 4
[2,] 2 5
[3,] 3 6

> c_list <- list(A,B)
> c_list

[[1]]
     a b
[1,] 1 4
[2,] 2 5
[3,] 3 6

[[2]]
  [,1] [,2] [,3]
a    1    2    3
b    4    5    6

> c_list <- c(c("one","two"),c_list)
> c_list

[[1]]
[1] "one"

[[2]]
[1] "two"

[[3]]
     a b
[1,] 1 4
[2,] 2 5
[3,] 3 6

[[4]]
  [,1] [,2] [,3]
a    1    2    3
b    4    5    6

> d_list <- list(c("one","two"),c_list)
> d_list

[[1]]
[1] "one" "two"
```

```
[[2]]
[[2]][[1]]
[1] "one"

[[2]][[2]]
[1] "two"

[[2]][[3]]
     a b
[1,] 1 4
[2,] 2 5
[3,] 3 6

[[2]][[4]]
  [,1] [,2] [,3]
a    1    2    3
b    4    5    6
```

Lists can be very useful for storing things, particularly different types of objects. Lists don't require the different elements to be of the same type. This feature makes them very useful and you will find them in the background for a number of **R** statistical functions. They can also store objects within objects.

One thing to notice above is that the operations c() and list() look similar but don't actually do the same thing. Look carefully at the difference.

B.2.4 Data Frames

Data frames are a very important object for statistical analysis. The data frame is like a matrix, but it allows different types of elements. The type must be the same for each column. Data frames also behave a lot like lists. Importantly, you can call a column using the $ symbol. See discussion in the next section.

```
> x <- read.csv("minimum wage.csv", as.is = TRUE)
> # as.is = TRUE keeps the data in the same format
> # as it was originally.
> # note R prefers to change types to factor.
> # another option is stringsAsFactors = FALSE.
> is.character(x$State)

[1] TRUE

> is.numeric(x$Minimum.Wage)

[1] TRUE
```

The object x is a data frame. It has one column of strings and two numeric columns.

B.3   Interacting with Objects

This section discusses how we can transform objects and retrieve information from them.

B.3.1   Transforming Objects

We can transform objects of various types into various other types with various degrees of success. Transformations, or confirmations, of an object type usually involve `as.someobjecttype()`. Note that `matrix` is often what I use to transform a vector into a matrix. Note that you need to state how many rows or columns it has.

```
> as.vector(B)

[1] 1 4 2 5 3 6

> matrix(c(a,b),nrow=3)

     [,1] [,2]
[1,]   1    4
[2,]   2    5
[3,]   3    6

> as.matrix(cbind(a,b))

     a b
[1,] 1 4
[2,] 2 5
[3,] 3 6

> as.list(B)

[[1]]
[1] 1

[[2]]
[1] 4

[[3]]
[1] 2

[[4]]
[1] 5
```

```
[[5]]
[1] 3

[[6]]
[1] 6

> unlist(a_list)

[1] 1 2 3 4 5 6

> as.character(A)

[1] "1" "2" "3" "4" "5" "6"

> as.factor(B)

[1] 1 4 2 5 3 6
Levels: 1 2 3 4 5 6

> as.vector(as.numeric(as.character(as.factor(a)))) == a

[1] TRUE TRUE TRUE
```

The oddest, and possibly the most frustrating type in **R**, is the **factor** type. **R** likes to store character vectors as factors because it is computationally efficient. But it is easy to confuse a factor object with a numeric object. They are not the same. I often have errors in my code due to this mistake.

B.3.2 Logical Expressions

Like other operations, logical operators are cell by cell in **R**. The == is used to determine whether something is true or false. We can also use order operations, >, <, >= and <=. The ! is used for not. The symbols & and | are used for "and" and "or" for combining logical arguments. You can also use && which will ask if *all* the elements are the same.

```
> a==b

[1] FALSE FALSE FALSE

> A==t(B)

         a    b
[1,] TRUE TRUE
[2,] TRUE TRUE
[3,] TRUE TRUE

> a_list[[1]]==a_list[[2]]
```

```
[1] FALSE FALSE FALSE
> a > b
[1] FALSE FALSE FALSE
> b > 5
[1] FALSE FALSE  TRUE
> b >= 5
[1] FALSE  TRUE  TRUE
> b <= 4
[1]  TRUE FALSE FALSE
> a != b
[1] TRUE TRUE TRUE
> (b > 4) & a == 3
[1] FALSE FALSE  TRUE
> (b > 4) && a == 3
[1] FALSE
> (b > 4) | a == 3
[1] FALSE  TRUE  TRUE
> (b > 4) || a == 3
[1] FALSE
```

B.3.3   Retrieving Information from a Position

There are a number of ways to retrieve information from objects. The simplest is to request it by its position in the object. Objects, such as vectors, have an index which gives the position of every object in the vector. Note **R** thinks of matrices as just rearranged vectors, so the index can also be used for matrices. As matrices are two-dimensional objects, information can also be retrieved using the matrix coordinates.

```
> a[1]
[1] 1
```

```
> b[3]

[1] 6

> A[5]

[1] 5

> A[2,2]

b
5
```

We are not limited to a single index number. We can retrieve a subset of the object using various index notation including - to mean "not." Note that there is no "end" index notation in **R**. In its stead I use length() or dim(). Note that dim() only works for objects with more than 1 dimension, like matrices.

```
> a[1:2]

[1] 1 2

> a[-3]

[1] 1 2

> A[1,]

a b
1 4

> A[c(1,5)]

[1] 1 5

> a[2:length(a)]

[1] 2 3

> D <- cbind(A,2*A)
> D

      a b a  b
[1,] 1 4 2  8
[2,] 2 5 4 10
[3,] 3 6 6 12

> D[,3:dim(D)[2]]
```

```
      a  b
[1,]  2  8
[2,]  4  10
[3,]  6  12
```

Retrieving information from lists can be frustrating. Consider the three different methods below. Note that the first two look very similar and seem to produce similar results, but they are actually quite different. If I understood the difference I would tell you! Suffice it to say, the double bracket thingy is probably what you want.

```
> a_list[2]

[[1]]
[1] 4 5 6

> a_list[2][2]

[[1]]
NULL

> a_list[[2]]

[1] 4 5 6

> a_list[[2]][2]

[1] 5
```

As well as the index, positions in an object may be named. In that case, the name can be used to retrieve the information. This is particularly useful for lists. To retrieve a named item in a list you can use the $. RStudio has a nice feature of popping up the options when you type the $, so you don't need to remember all the names.

```
> names(a_list) <- c("first", "second")
> a_list$first

[1] 1 2 3

> names(a_list)

[1] "first"   "second"

> names(a_list)[2]

[1] "second"

> b_list <- list(a_list,B=B)
> b_list
```

```
[[1]]
[[1]]$first
[1] 1 2 3

[[1]]$second
[1] 4 5 6


$B
   [,1] [,2] [,3]
a    1    2    3
b    4    5    6

> b_list$B

   [,1] [,2] [,3]
a    1    2    3
b    4    5    6
```

B.3.4 Retrieving the Position from the Information

Often we are trying to determine where certain information lies in the object. We can use logical expressions and the function which() to find the index positions of vectors.

```
> which(a > 2)

[1] 3

> which(A > 2)

[1] 3 4 5 6

> which(a_list[[2]] > 2)

[1] 1 2 3
```

If we are going to use the index to subset another object, we don't need to use which().

```
> b[a > 2]

[1] 6

> B[A > 2]

[1] 2 5 3 6
```

```
> A[,colnames(A)=="b"]
```

```
[1] 4 5 6
```

We can also use `match()` or `%in%` to determine which information belongs to a set. Note that `%in%` returns `TRUE` or `FALSE`, while `match()` returns the index or `NA` if there is no match. We can use `match()` to find objects in a list. Note that we need to be careful that we are looking for the correct type of object in the list. But `match()` can also find things that are not exactly format. See example below.

```
> d <- c("one", "two", "three", "four")
> c("one", "four", "five") %in% d
```

```
[1]    TRUE   TRUE FALSE
```

```
> match(a,c(1,2))
```

```
[1]   1   2 NA
```

```
> match(a_list,c(4,5,6))
```

```
[1] NA NA
```

```
> match(a_list,list(c(4,5,6)))
```

```
[1] NA   1
```

```
> match("1",c(3,5,8,1,99))
```

```
[1] 4
```

To find a word or part of a string we can use `grep()` and the associated functions. These functions look for patterns in object and create an index for where the pattern is true. Note that these can also be used to determine if the pattern is in the object at all. These functions can use **real expressions**, which I find so unintuitive that I generally avoid. More generally, I don't think **R** does a great job at analyzing strings. I often find it is easier to switch to Python for text analysis.

```
> grep("two", d)
```

```
[1] 2
```

B.4   Statistics

This section discusses basic statistical operations in **R**.

B.4.1 Data

```
> set.seed(123456789)
> a <- c(1:1000)
> b <- c("one","two",NA,4,5:1000)
> e <- rnorm(1000)
> c <- 2 - 3*a + e
> x <- as.data.frame(cbind(a,b,c))
```

Consider the simulated data. The data frame object has a nice property that allows you to have objects in which some variables are numeric while others are strings. That said, you need to be careful and keep track of the variable types. Note that all the variables in x are factors. **R** has transformed them.

```
> x <- as.data.frame(cbind(a,b,c), stringsAsFactors = FALSE)
```

What type are the variables in x now?

```
> x$a <- as.numeric(x$a)
> x$c <- as.numeric(x$c)
> write.csv(x,"x.csv")
```

Some useful functions for creating simulated data are rnorm() and runif(). These are random number generators. You can generate **pseudo-random numbers** from various **R** functions. Use related functions to calculate probability distributions and densities. Note that these numbers are generated according to a particular function. They are generated such that if they start with a particular number, *a seed*, then they will always give the same sequence. So if you can get the system to always start with the same number, you can have the results be exactly reproducible. In **R**, set.seed() does this. I always set the seed with 123456789.

As a general rule, I like to keep my scripts in the same folder as the data. In this way there are no long paths for writing or reading files. If you want to do this in **RStudio**, then make sure to go to Session > Set Working Directory > To Source File Location.

B.4.2 Missing Values

R has a particular way of dealing with missing values. It uses NA for missing. Any operation of a value with a missing will give back a missing. This is a nice feature which forces the programmer to think about how to deal with missing information. **R** will also treat NA as numeric.

```
> 2 + NA
```

```
[1] NA
```

```
> 2*NA
```

```
[1] NA
```

```
> 2 + c(3,4,6,7,NA)
```

```
[1]  5  6  8  9 NA
```

Often in microeconometrics we want to ignore the missing values. We may be assuming that they are **missing-at-random**. We did this in our analysis of returns to schooling for men in Chapters 1 and 2.

We can have many **R** functions skip over the missing using the option `na.rm = TRUE`. We see this option used below. Some functions, like `lm()`, will do this automatically.

B.4.3 Summary Statistics

We can do some basic summary statistics on the data using `mean()` and `sd()`. We can also use the quantile function to find out more about the distribution.

```
> mean(x$a)
```

```
[1] 500.5
```

```
> sd(x$a)
```

```
[1] 288.8194
```

```
> quantile(x$a,c(1:5)/5)
```

```
   20%    40%    60%    80%   100%
 200.8  400.6  600.4  800.2 1000.0
```

```
> # rowMeans(x[,c(1,3)])
> colMeans(x[,c(1,3)])
```

```
        a          c
  500.500 -1499.488
```

```
> x$d <- NA
> x[2:1000,]$d <- c(2:1000)/10
> # mean(x$d)
> mean(x$d, na.rm = TRUE)
```

```
[1] 50.1
```

B.4.4 Regression

The standard OLS function is `lm()` for linear model. When this function is run it creates a list of objects which can be used for various things. For example, you can recall just the coefficient estimates or the residuals from the regression. A nice feature of these functions is that you can just use the variable names in the regression call. You can do this as long as you specify the data frame object using `data=`. There are various built-in non-linear regression functions that may be called using the `glm()` procedure.

```
> x <- read.csv("x.csv", as.is = TRUE)
> #summary(x)
> lm1 <- lm(c ~ a, data=x)
> summary(lm1)[[4]]

            Estimate    Std. Error      t value        Pr(>|t|)
(Intercept) 2.097396 0.0641312421      32.70475 4.934713e-160
a          -3.000170 0.0001109953 -27029.69905   0.000000e+00

> lm1$coefficients

(Intercept)          a
   2.097396   -3.000170

> glm1 <- glm(c > -1500 ~ a, family = binomial(link=probit),
+             data=x)
> glm1$coefficients

(Intercept)          a
3799.867658   -7.592143
```

Note that when **R** reads in data using the `read.csv()` function it creates an extra variable.

B.5 Control

This section discusses basic control methods used in **R**. There is a folklore that you cannot do loops in **R**. This is not quite correct. There are good and bad ways to do loops in **R**.

B.5.1 Loops

Looping is a fundamental part of programming. The basic loop is the "for-loop." The "while-loop" is an alternative. The `for ()` loop is better when there is a particular number of iterations, while the `while ()` loop is better when you want it to stop when some value is reached.

B.5.2 Looping in R

Looping in **R** has special challenges. Consider the following problem of creating
a matrix with three consecutive numbers in each row.

```
> # don't do it this way!!!
> start_time <- Sys.time()
> A <- NULL
> for (i in 1:10000) {
+    A <- rbind(A,c(i,i+1,i+2))
+ }
> Sys.time() - start_time

Time difference of 2.079877 secs

> A[400,]

[1] 400 401 402

> sum(A)

[1] 150045000

> # A faster way
> start_time <- Sys.time()
> A <- matrix(NA,10000,3)
> for (i in 1:10000) {
+    A[i,] <- c(i,i+1,i+2)
+ }
> Sys.time() - start_time

Time difference of 0.025877 secs

> A[400,]

[1] 400 401 402

> sum(A)

[1] 150045000

> # An even faster way! (Sometimes)
> start_time <- Sys.time()
> A <- t(matrix(sapply(1:10000,
+                      function(x) c(x,x+1,x+2)),nrow = 3))
> Sys.time() - start_time

Time difference of 0.03279209 secs
```

```
> A[400,]
```

```
[1] 400 401 402
```

```
> sum(A)
```

```
[1] 150045000
```

There is a particular way *not* to loop in **R**. **R** loops are very very slow if they involve creating and recreating objects. These types of loops also tend to be very memory hungry. In the previous example you can get a significant speed increase by creating the object at the start and then filling it in during the loop.

The example above also illustrates the `apply()` function. Using `apply()` is often the fastest way of looping. I don't find it very intuitive and so I tend not to use this method, but it may be useful if you are trying to speed up some code.

B.5.3 If Else

R uses two different types of "if-else" commands. The command you will use most of the time is `ifelse()`. This function operates on vectors. The command takes three arguments. The first is the logical statement to be checked, the second is what to do if the statement is true and the third is what to do if the statement is false. Note that like other operations on vectors this function works cell by cell.

```
> a <- c(1,2,3,4,5)
> b <- ifelse(a==3,82,a)
> a
```

```
[1] 1 2 3 4 5
```

```
> b
```

```
[1]  1  2 82  4  5
```

The more standard programming command is `if ()` or `if () else`. This command checks the logical statement and then runs the command in the curly brackets. Note that you don't have to specify the "else."

```
> A <- "Chris"
> # A <- "Write Your Name Here"
> if (A=="Chris") {
+   print("Hey Chris")
+ } else {
+   print(paste("Hey",A))
+ }
```

```
[1] "Hey Chris"
```

B.6 Optimization

Optimization is fundamental to statistical modeling. The main command in **R** uses functions as inputs, so the section discusses how to create and use these objects. It also discusses the basic optimizer, optim().

B.6.1 Functions

If you plan to have procedures that you will use repeatedly, then it may be more efficient to create a function. Functions also help keep your script clean and tidy. They also make debugging easier by focusing you where the location of the bug is.

```
> y <- x[,c(2,4)]
> apply(y, 2, mean)

        a         c
 500.500 -1499.488

> colMeans(y)

        a         c
 500.500 -1499.488

> b <- c(1:dim(y)[1])
> summary(sapply(b, function(x) sum(y[x,])), digits = 2)

   Min. 1st Qu.  Median    Mean 3rd Qu.    Max.
-2000.0 -1500.0 -1000.0 -1000.0  -500.0     0.5

> summary(rowSums(y), digits = 2)

   Min. 1st Qu.  Median    Mean 3rd Qu.    Max.
-2000.0 -1500.0 -1000.0 -1000.0  -500.0     0.5
```

We can define functions on the fly like when we use the apply() function.

```
> my_mean <- function(x) {
+   if (is.numeric(x)) {
+     return(mean(x, na.rm = TRUE))
+   }
+   else return("Not Numeric!")
+ }
> my_mean(x$b)

[1] "Not Numeric!"
```

```
> my_mean(x$a)

[1] 500.5

> my_mean(c(x$a,NA,mean(x$a)))

[1] 500.5
```

When writing functions it is good practice to check that the input is the correct type and provide a warning or error code if it is not the expected type.

```
> lm_iv <- function(y_in, X_in, Z_in = X_in, Reps = 100,
+                    min_in = 0.05, max_in = 0.95) {
+   # takes in the y variable, x explanatory variables
+   # and the z variables if available.
+   # defaults: Z_in = X_in,
+   # Reps = 100, min_in = 0.05, max_in = 0.95
+
+   # Set up
+   set.seed(123456789)
+   index_na <- is.na(rowSums(cbind(y_in,X_in,Z_in)))
+   yt <- as.matrix(y_in[index_na==0])
+   Xt <- as.matrix(cbind(1,X_in))
+   Xt <- Xt[index_na==0,]
+   Zt <- as.matrix(cbind(1,Z_in))
+   Zt <- Zt[index_na==0,]
+   N_temp <- length(yt)
+   # turns the inputs into matrices
+   # removes observations with any missing values
+   # add column of 1s to X and Z
+
+   # Bootstrap
+   r <- c(1:Reps)
+   bs_temp <- sapply(r, function(x) {
+     ibs <- round(runif(N_temp, min = 1, max = N_temp))
+     solve(t(Zt[ibs,])%*%Xt[ibs,])%*%t(Zt[ibs,])%*%yt[ibs]
+   } )
+
+
+   # Present results
+   res_temp <- matrix(NA,dim(Xt)[2],4)
+   res_temp[,1] <- rowMeans(bs_temp)
+   for (j in 1:dim(Xt)[2]) {
+     res_temp[j,2] <- sd(bs_temp[j,])
+     res_temp[j,3] <- quantile(bs_temp[j,],min_in)
+     res_temp[j,4] <- quantile(bs_temp[j,],max_in)
```

```
+   }
+   colnames(res_temp) <-
+     c("coef","sd",as.character(min_in),as.character(max_in))
+   return(res_temp)
+ }
```

Here is a version of the instrumental variable function. This function uses matrix algebra to calculate the instrumental variable regression. It presents the bootstrap results. The input variables listed at the start of the function can be set to default values. This means that the user doesn't have to enter these values. At the start of the function you can define objects and variables. This is a good place to confirm that objects are what you think they are. This version of the function uses `sapply()` to do the bootstrap loop.[2]

Note also the use of **global** and **local variables** in **R**. A function can define variables that are local to that function. That is a variable with the same name can have two different values, one outside the function and one inside the function. To send that information back, you can define global variables that call the function. Global variables can be used by a function without being defined within the function. However, I would generally try to avoid doing this as it can lead to errors and confusion. It is also a good idea to use separate names for local variables, so as not to get confused. In the code above `_in`, `t` and `_temp` are used to denote local or temporary variables.

B.6.2   optim()

The standard optimization procedure in **R** is `optim()`. The algorithms used in **R** were developed by John C. Nash (no relation to the Nobel prize winning mathematician) in the 1970s. Nash developed the optimizer in a language called BASIC in order to run on the particular machine that Nash was using. According to John C. Nash, there are better optimization algorithms available (Nash, 2014).

```
> f_ols <- function(beta, y_in, X_in) {
+   X_in <- as.matrix(cbind(1,X_in))
+   if (length(beta)==dim(X_in)[2]) {
+     return(mean((y_in - X_in%*%beta)^2, na.rm = TRUE))
+   }
+   else {
+     return("The number of parameters doesn't match.")
+   }
+ }
> lm_ols <- optim(par=c(2,-3),fn=f_ols,y_in=x$c,X_in=x$a)
> lm_ols
```

---

[2]In some cases it may be faster to call a compiled language like C. It is also possible to create compiled functions in **R**, see Kabacoff (2011).

```
$par
[1]   2.097287 -3.000170

$value
[1] 1.024609

$counts
function gradient
      67       NA

$convergence
[1] 0

$message
NULL

> lm(x$c ~ x$a)

Call:
lm(formula = x$c ~ x$a)

Coefficients:
(Intercept)          x$a
      2.097       -3.000
```

Above is a simple least squares function. The problem is solved using `optim()`. The function, `optim()`, takes initial starting values, a function to optimize and variable values. It has a large number of default settings. In particular, it uses the Nelder-Mead procedure as a default. One thing to note is that this procedure may give different results depending on the starting values used. What happens if you change the starting values to `par=c(1,1)`?

B.7 Discussion and Further Reading

While I have been programming for a very long time, I am not a programmer. The objective of this chapter is to give you some basic insights into **R**. If you are interested in really learning how to program in **R** then I suggest purchasing one of the many books out there or taking an online course. I highly recommend Kabacoff (2011) as it is written by a statistician/computer scientist. I also suggest Paarsh and Golyaev (2016) as an excellent introduction to data analysis in economics.

B

Note

The views expressed here are those of the author and do not reflect those of the Federal Trade Commission or the Congressional Budget Office.

I am grateful for the help I received from friends and colleagues who read and commented on the book. I am particularly grateful to Emek Basker, David Vanness and Greg Kirwin who read early versions of the early chapters. I am grateful to George Deltas, Peter Newberry, Joris Pinkse and Nathan Wilson, for their thoughts and insights into the "structural chapters." Also, to Ott Toomet and Grant Farnsworth who reviewed sections at the request of the editor. I am very grateful to Brian Krauth, David Prentice and Devesh Raval who worked through the whole text making suggestions and edits. Finally, I am grateful to CJ Adams for helping work through the code.

Thank you to the editing team, particularly David Grubbs, Lara Spieker, Kari Budyk and Shashi Kumar.

All errors are my own.

Bibliography

Alberto Abadie, Alexis Diamond, and Jens Hainmueller. Synthetic control methods for comparative case studies: estimating the effect of California's tobacco control program. *Journal of the American Statistical Association*, 105(490):493–505, 2010.

Christopher P. Adams. Finite mixture models with one exclusion restriction. *The Econometrics Journal*, 19:150–165, 2016.

Christopher P. Adams. Measuring treatment effects with big n panels. Federal Trade Commission, July 2018.

Abhay Aneja, John J. Donohue III, and Alexandria Zhang. The Impact of Right-to-Carry Laws and the NRC Report: Lessons for the Empirical Evaluation of Law and Policy. *American Law and Economics Review*, 13(2): 565–632, 2011.

Joshua D. Angrist and Alan B. Krueger. Instrumental variables and the search for identification: From supply and demand to natural experiments. *Journal of Economic Perspectives*, 15(4):69–85, 2001.

Gaurab Aryal, Serafin Grundl, Dong-Hyuk Kim, and Yu Zhu. Empirical relevance of ambiguity in first-price auctions. *Journal of Econometrics*, 204 (2):189–206, 2018.

Orley Ashenfelter and Alan Krueger. Estimates of the economic return to schooling from a new sample of twins. *American Economic Review*, 84(5), 1994.

Orley Ashenfelter and Cecilia Rouse. Income, schooling, and ability: Evidence from a new sample of identical twins. *Quarterly Journal of Economics*, 113 (1):253–284, 1998.

Nava Ashraf, Dean Karlan, and Wesley Yin. Tying odysseseus to the mast: Evidence from a commitment savings product in the philippines. *The Quarterly Journal of Economics*, 121(2):635–672, May 2006.

John Asker. A study of the internal organization of a bidding cartel. *American Economic Review*, 100(3):724–762, 2010.

Susan Athey and Philip A. Haile. Identification in standard auction models. *Econometrica*, 70(6):2170–2140, 2002.

Susan Athey, Jonathan Levin, and Enrique Seira. Comparing open and sealed bid auctions: Evidence from timber auctions. *The Quarterly Journal of Economics*, 126:207–257, 2011.

Susan Athey, Mohsen Bayati, Guido Imbens, and Zhaonan Qu. Ensemble methods for causal effects in panel data settings. *American Economics Review Papers and Proceedings*, 109, May 2019.

Jushan Bai. Panel data models with interactive fixed effects. *Econometrica*, 77:1229–1279, 2009.

Laura H. Baldwin, Robert C. Marshall, and Jean-Francois Richard. Bidder collusion at forest service timber sales. *Journal of Political Economy*, 105 (5):657–699, 1997.

Tatiana Benaglia, Didier Chauveau, and David R. Hunter. An em-like algorithm for semi- and non-parametric estimation in multivariate mixtures. *Journal of Computational and Graphical Statistics*, 18(2):505–526, 2009.

James O. Berger. *Statistical Decision Theory and Bayesian Analysis*. Springer, 1985.

Steven T. Berry. Estimating discrete-choice models of product differentiation. *RAND Journal of Economics*, 25(2):242–262, 1994.

Steven T. Berry, James Levinsohn, and Ariel Pakes. Automobile prices in market equilibrium. *Econometrica*, 60(4):889–917, 1995.

Steven T. Berry, Amit Gandhi, and Philip A. Haile. Connected substitutes and invertibility of demand. *Econometrica*, 81:2087–2111, 2013.

David M. Blei. Probabilistic topic models. *Communications of ACM*, 55(4): 77–84, April 2012.

Tim Bresnahan. The Apple-Cinnamon Cheerios war: valuing new goods, identifying market power, and economic measurement. Stanford University, 1997.

A. Colin Cameron and Pravin K. Trivedi. *Microconometrics: Methods and Applications*. Cambridge University Press, 2005.

David Card. *Aspects of Labour Market Behavior: Essays in Honour of John Vanderkamp*, chapter Using Geographic Variation in College Proximity to Estimate the Return to Schooling, pages 201–222. University of Toronto Press, 1995.

David Card. Estimating the returns to schooling: Progress on some persistent econometric problems. *Econometrica*, 69(5):1127–1160, September 2001.

David Card and Alan B. Krueger. Minimum wages and employment: a case study of the fast-food industry in New Jersey and Pennsylvania. *The American Economic Review*, 84(4):772–793, September 1994.

Janet Currie and Bruce C. Fallick. The minimum wage and the employment of youth: evidence from the NLSY. *Journal of Human Resources*, 31(2): 404–424, 1996. Spring.

George Deltas. Asymptotic and small sample analysis of the stochastic properties and certainty equivalents of winning bids in independent private values auctions. *Economic Theory*, 23:715–738, 2004.

Nikolay Doudchenko and Guido W. Imbens. Balancing, Regression, Difference-in-Difference and Synthetic Control Methods: A synthesis. NBER Working Paper 22791, October 2016.

Bradley Efron. Two modeling strategies for empirical Bayes estimation. *Statistical Science*, 29:285–301, 2014.

Kohei Enami and John Mullahy. Tobit at fifty: a brief history of Tobin's remarkable estimator, of related empirical methods, and of limited dependent variable econometrics in health economics. *Health Economics*, 18(6):619–628, 2009.

Michal Fabinger and Glen Weyl. Pass-through as an economic tool: Principle of incidence under imperfect competition. *Journal of Political Economy*, 121 (3):528–583, 2013.

Yanqin Fan and Sang Soo Park. Sharp bounds on the distribution of treatment effects and their statistical inference. *Econometric Theory*, 26(3):931–951, 2010.

Xiao Fu, Kejun Huang, Bo Yang, and Nicholas D. Sidiropoulos. Robust volume minimization-based matrix factorization for remote sensing and document clustering. *IEEE Transactions on Signal Processing*, 64(23):6254–6268, December 2016.

Itzhak Gilboa. Lecture Notes for Introduction to Decision Theory. Yale, March 2013.

Arthur Goldberger. *A Course in Econometrics*. Harvard University Press, 1991.

William Greene. *Econometric Analysis*. Prentice Hall, fourth edition, 2000.

Emmanuel Guerre, Isabelle Perrigne, and Quang Vuong. Optimal nonparametric estimation of first-price auctions. *Econometrica*, 68(3):525–574, 2000.

Philip A. Haile and Yuichi Kitamura. Unobserved heterogeneity in auctions. *The Econometrics Journal*, 22:C1–C19, 2019.

Philip A. Haile and Elie T. Tamer. Inference with incomplete model of English auctions. *Journal of Political Economy*, 111(1), 2003.

Philip A. Haile, Han Hong, and Matt Shum. Nonparametric tests for common values in first-price sealed-bid auctions. Yale, November 2006.

Lars Peter Hansen. Large sample properties of generalized method of moments estimators. *Econometrica*, 50(4):1029–1054, 1982.

Jerry Hausman. *The Economics of New Goods*, chapter Valuation of New Goods under Perfect and Imperfect Competition. Number 58 in NBER Studies in Income and Wealth Number. The University of Chicago Press, 1997.

James Heckman and Bo Honore. The empirical content of the Roy model. *Econometrica*, 58:1128–1149, 1990.

Marc Henry, Ismael Mourifie, and Romuald Meango. Sharp bounds and testability of a Roy model of STEM major choices. *Journal of Political Economy*, 2020. forthcoming.

Marek Hlavac. stargazer: Well-formatted regression and summary statistics tables. R package version 5.2.2, 2018.

Benjamin Hoffman. For the Sultan of Small Sample Size, a 1.000 Career Average. *The New York Times*, 10(1), March 2013.

Harold Hotelling. Stability in competition. *The Economic Journal*, 39(153): 41–57, March 1929.

Harold Hotelling. Analysis of a complex of statistical variables into principal components. *Journal of Educational Psychology*, 24(6):417–441, September 1933. http://dx.doi.org/10.1037/h0071325.

Yingyao Hu. The econometrics of unobservables: Applications of measurement error models in empirical industrial organization and labor economics. *Journal of Econometrics*, 200(2):154–168, October 2017.

Kejun Huang, Nicholas D. Sidiropoulos, and Ananthram Swami. Non-negative matrix factorization revisited: Uniqueness and algorithm for symmetric decomposition. *IEEE Transactions on Signal Processing*, 62(1), 2014.

Guido W. Imbens. Better LATE than nothing: some comments on Deaton (2009) and Heckman and Urzua (2009). *Journal of Economic Literature*, 48 (2):399–423, June 2010.

Sonia Jaffe and Glen Weyl. The first-order approach to merger analysis. *American Economic Journal: Microeconomics*, 5(4):188–218, 2013.

Robert I. Kabacoff. *R in Action*. Manning, 2011.

Désiré Kédagni. Identifying Treatment Effects in the Presence of Confounded Types. Working paper from the Iowa State University, 2017.

Christian Kleiber and Achim Zeileis. *Applied Econometrics with R (Use R!)*. Springer, 2008.

Peter Klibanoff, Massimo Marinacci, and Sujoy Mukerji. A smooth model of decision making under ambiguity. *Econometrica*, 73(6):1849–1892, November 2005.

Elena Krasnokutskaya. Identification and estimation in procurement auctions under unobserved auction heterogeneity. *Review of Economic Studies*, 78(1): 293–327, 2011.

David Kreps. *A Course in Microeconomic Theory*. Harvester Wheatsheaf, 1990.

Michael Levine, David R. Hunter, and Didier Chauveau. Maximum smoothed likelihood for multivariate mixtures. *Biometrika*, 98(2):403–416, 2011.

Alexander MacKay and Nathan Miller. Estimating models of supply and demand: Instruments and covariance restrictions. Harvard Business School, October 2019.

Charles F. Manski. Nonparametric bounds on treatment effects. *American Economic Review Papers and Proceedings*, 80:319–323, 1990.

Charles F. Manski. The lure of incredible certitude. *Economics and Philosophy*, 36(2):216–245, 2020.

Charles F. Manski and John V. Pepper. Deterrence and the death penalty: Partial identification analysis using repeated cross sections. *Journal of Quantitative Criminology*, 29:123–141, 2013.

Charles F. Manski and John V. Pepper. How do right-to-carry laws affect crime rates? coping with ambiguity using bounded-variation assumptions. *Review of Economics and Statistics*, 100(2):232–244, 2018.

Charlies F. Manski. *Analog Estimation Methods in Econometrics*. Chapman and Hall, 1988.

Charlies F. Manski. *Identification Problems in the Social Sciences*. Harvard University Press, 1995.

Daniel McFadden. The measurement of urban travel demand. *Journal of Public Economics*, 3:303–328, 1974.

John Mullahy. Individual Results May Vary: Inequality-probability bounds for some health-outcome treatment effects. *Journal of Health Economics*, 61: 151–162, 2018.

362Bibliography

Alicia H. Munnell, Geoffrey M. B. Tootell, Lynn E. Brown, and McEneaney. Mortgage lending in Boston: interpreting HMDA data. *The American Economic Review*, 86(1):25–53, March 1996.

John C. Nash. On best practice optimization methods in **R**. *Journal of Statistical Software*, 60(2), 2014.

Aviv Nevo. A practioner's guide to estimation of random-coefficients logit models of demand. *Journal of Economics and Management Strategy*, 9(4): 513–548, 2000.

Alexei Onatski and Chen Wang. Spurious factor analysis. *Econometrica*, 2020. Forthcoming.

Harry J. Paarsh and Konstantin Golyaev. *A Gentle Introduction to Effective Computing in Quantitative Research: What Every Research Assistant Should Know.* MIT Press, 2016.

Ariel Pakes, Jack Porter, Kate Ho, and Joy Ishii. Moment inequalities and their application. *Econometrica*, 83(1):315–334, 2015.

Judea Pearl and Dana Mackenzie. *The Book of Why: The New Science of Cause and Effect.* Basic Books, 2018.

Herbert Robbins. The empirical Bayes approach to statistical decision problems. *Proceedings of the Third Berkeley Symposium on Mathematical Statistics and Probability*, pages 157–163, 1956. University of California Press.

David Romer. Do firms maximize? Evidence from professional football. *Journal of Poltical Economy*, 114(2):340–365, April 2006.

Donald B. Rubin. Estimating causal effects of treatments in randomized and nonrandomized studies. *Journal of Educational Psychology*, 66(5), 1974.

James H. Stock and Mark W. Watson. *Introduction to Econometrics.* Pearson, third edition, 2011.

Robert Tibshirani. Regression shrinkage and selection via the lasso. *Journal of the Royal Statistical Society. Series B (Methodological)*, 58(1):267–288, 1996.

Hal Varian. Position auctions. *International Journal of Industrial Organization*, 25(6):1163–1178, December 2007.

Abraham Wald. Statistical decision functions. *Annals of Mathematical Statistics*, 20(2):165–205, 1949.

Author index

Subject index

wide data problem, 255–257, 259,
        261

zero-sum game, 185, 187

Printed in the United States
By Bookmasters